Zoophysiology   Volume 25

Zoophysiology

G. D. Pollak   J. H. Casseday

The Neural Basis
of Echolocation in Bats

With 82 Figures

Springer-Verlag
Berlin Heidelberg New York
London Paris Tokyo

Professor Dr. GEORGE D. POLLAK
Department of Zoology
The University of Texas at Austin
Austin, Texas 78712, USA

Professor Dr. JOHN H. CASSEDAY
Departments of Surgery and Neurobiology
Duke University Medical Center
Durham, North Carolina 27710, USA

QL
737
.C5
P65
1989

Cover illustration: Pteronotus parnellii drawn after a photograph by Ellen Covey

ISBN 3-540-50520-2 Springer-Verlag Berlin Heidelberg New York
ISBN 0-387-50520-2 Springer-Verlag New York Berlin Heidelberg

Library of Congress Cataloging-in-Publication Data. Pollak, G. D. (George D.),
1942– . The neural basis of echolocation in bats/G. D. Pollak, J. H. Casseday. p.
cm. – (Zoophysiology: v. 25) Bibliography: p. Includes index. 1. Bats – Physiology.
2. Echolocation (Physiology). 3. Mammals – Physiology. I. Casseday, J. H. (John H.),
1934– . II. Title. III. Series. QL737.C5P65 1989 599.4'01852–dc 19 89-4081

Typesetting, printing and binding: Brühlsche Universitätsdruckerei, Giessen
2131/3145-543210 – Printed on acid-free paper

Preface

The brain of an echolocating bat is devoted, in large part, to analyzing sound and conducting behavior in a world of sounds and echoes. This monograph is about analysis of sound in the brainstem of echolocating bats and concerns the relationship between brain structure and brain function.

Echolocating bats are unique subjects for the study of such relationships. Like man, echolocating bats emit sounds just for the purpose of listening to them. Simply by observing the bat's echolocation sounds, we know what the bat listens to in nature. We therefore have a good idea what the bat's auditory brain is designed to do. But this alone does not make the bat unique. The brain of the bat is, by mammalian standards, rather primitive. The unique aspect is the combination of primitive characteristics and complex auditory processing. Within this small brain the auditory structures are hypertrophied and have an elegance of organization not seen in other mammals. It is as if the auditory pathways had evolved while the rest of the brain remained evolutionary quiescent.

The study of physiological mechanisms for echolocation has a long history, part of which we review here. Progress in the study of the structural mechanisms for analyzing echoes, mainly within the last decade, was a major stimulus for writing this monograph. Our aim was quite simple. We thought there is now enough known about the function *and* structure of the bat's central auditory pathways that we could provide a synthesis, a structure-function story. Of course the story has gaps which we have tried to point out as incentives for further research.

The ideas expressed in this monograph were obtained in large part from our own research experience, but they were also profoundly influenced by our teachers, students and colleagues. We wish to acknowledge our gratitude to these scientists for enriching our perspectives and providing encouragement, sometimes during difficult periods. We offer special thanks to Gerhard Neuweiler. We are indebted to his unfailing encouragement, for the high scientific standards he has set for us all, and for the intellectually rigorous yet romantic insights that he has given us into the functioning of the auditory system. We also thank O. W. Henson, Jr., both for teaching us about

V

the world of bats and for his generous suggestions for pursuing investigations of their auditory systems. Our ideas were also strongly influenced by both Donald Griffin and Irving Diamond. Professor Griffin discovered and developed the field of echolocation, and he encouraged all who would listen to pursue studies of the bat's nervous system, even when such studies were viewed by many as esoteric endeavors. Professor Diamond provided training in comparative neuroanatomy, not only for one of us (JHC) but for a host of other neuroanatomists. His influence is strongly imprinted on the views presented here.

During the writing of this monograph a number of scientists provided valuable comments, suggestions and discussions. These people include J. J. Blum, Ellen Covey, Nick Fuzessary, Gerhard Neuweiler, Tom Parisi, Linda Ross, Wesley Thompson, Harold Zakon and John Zook. Finally we express our appreciation for the financial support of the National Science Foundation and the National Institutes of Health. Our research and the writing of this monograph was supported by grants BNS 85-20441 (JHC), NS 21748 (JHC) and NS 21286 (GDP).

January 1989 GEORGE D. POLLAK
JOHN H. CASSEDAY

Contents

Photograph of a flying mustache bat chasing a moth, reproduced with kind permission of the photographer, Mr. Russell Hansen

Chapter 1

Biological Sonar and the World of Bats

1.1 Purpose of the Monograph

The study of echolocating bats represents a major triumph in attempts to elucidate the neural basis of behavior. An elegant and comprehensive picture has emerged. It encompasses the entire auditory system from cochlea to cortex, and shows many of the mechanisms that allow these animals to form percepts of the external world through their sense of hearing.

The development of this research followed a natural evolution and can be divided into two major periods. The first period began in 1963 with the publication of the initial neurophysiological studies of the bats' auditory system by Alan Grinnell (1963a, b, c, d). In the following year the first of Nobuo Suga's studies appeared (Suga 1964a, b), as did Frishkopf's (1964) study of the auditory nerve of the little brown bat. Interest in the neural basis of echolocation then quickly spread to other laboratories. By the early 1970's there was a rich literature describing the special adaptations of echolocating bats. Some studies addressed binaural processing and a few addressed more directly questions about the mechanisms underlying the responses elicited by the simulated biosonar signals. However, the research produced during this period was dominated by the recording of audiograms and the demonstration that the bat's nervous system was specially adapted to respond to a faint echo very shortly after the emission of a loud orientation call (e.g., Suga 1964a, 1967; Friend et al. 1966; Grinnell 1967, 1970, 1973; Henson 1967; Neuweiler 1970; Grinnell and Hagiwara 1972). These data, obtained from different bat species from all over the world, showed that each species was most sensitive to the frequency spectrum of its own orientation call and documented the remarkable abilities to respond to two signals following in close temporal sequence. An intriguing feature was that the audiogram of each species had its own special adaptations. These adaptations suggested that each species has some novel mechanisms and challenged further investigation. The adaptations seen in the audiograms of the mustache bat (Grinnell 1967; Pollak et al. 1972) and the horseshoe bat (Neuweiler 1970) were espicially pronounced and attracted considerable attention from a number of investigators.

In 1975 and 1976 two sets of publications signaled the beginning of the second period. One report, from Nobuo Suga's laboratory, dealt with the peripheral auditory system of the mustache bat (Suga et al. 1975). Four other reports were from the group at Frankfurt University, where Suga was at the time a visiting scientist, and concerned the peripheral auditory system of the horseshoe bat (Bruns 1976a, b; Schnitzler et al. 1976; Suga et al. 1976). Several aspects of these reports set them apart from previous studies. For the first time the frequency representation of a bat's cochlea was mapped and the marked morphological features of the basilar membrane were described. More importantly, however, these studies established a direct relationship between structure and function in the bat's auditory system, where the anatomical and physiological features of the cochlea were causally linked to the specialized properties of auditory nerve fibers. These features, in turn, were closely correlated with the auditory abilities of these animals. Whereas the emphasis of previous studies was on demonstrating the existence of special adaptations, the elegance and clarity of these studies showed that the adaptations can be understood on a mechanistic level, and they gave notice that the two species were to be the subjects of intense future investigations.

The period around 1976 was also notable because it marked the re-emergence of neuroanatomical studies of the central auditory pathway of bats. Modern methods were applied to the study of the architecture and connections in the auditory pathway of mustache bats at Duke University and of horseshoe bats at Frankfurt University. Historically, the remarkable hypertrophy of the central auditory pathways in bats was recognized long before the biological sonar of these animals was discovered. Some of the first experimental studies of the auditory pathways were conducted in 1926 by Poljak on horseshoe bats, and it seems paradoxical that after the discovery of echolocation, studies of the connections of the central auditory pathway were neglected for more than 30 years.

The progress made over the past 10 years has been remarkable by any standard. Studies of the cochlea, cochlear nucleus, superior olive, inferior colliculus and cortex have provided major insights into the processing of biosonar signals and how that information is represented in the brain by ensembles of neurons. The ideas and techniques of "bat" neurophysiologists have blended naturally with those of comparative neuroanatomists, and have led to the recognition of homologies between structures in the auditory systems of less specialized animals and the auditory systems of bats. This recognition makes us think somewhat more broadly about both the physiological and anatomical features that emerge from studies of bats. The emphasis during the early period was to demonstrate neural correlates of echolocation and to think of those correlates as some sort of special adaptation. The studies over the past decade permit a deeper understanding of the adaptations, thus allowing investigators to specify how and why the special adaptations represent exaggerated features of a basic design present in all mammals.

In view of this substantial progress, we felt it an appropriate time to present an integrated picture of the chiropteran auditory system, from periphery to midbrain, based largely on studies that appeared since 1975. We will not consider the studies of the bat's auditory cortex since this topic has been treated in several previous reviews (Suga 1978, 1984, 1988; Suga and O'Neill 1980). Our basic premise is that the adaptations evolved from a basic mammalian auditory system to confer a competitive advantage for process-

4

ing some feature of the acoustic signal. These adaptations are useful to neuroscientists because they present architectural and mechanistic features of the mammalian auditory system in exaggerated form. Thus the monograph will focus on how a number of basic organizational principles are expressed as special adaptations in the auditory systems of bats.

Below we give a brief preview of the topics and issues that we address in the subsequent sections and chapters. We begin in the following section with a brief historical account of the discovery of echolocation and describe the structure of the biosonar signals bats use for orientation. We then briefly discuss two major types of biosonar systems and the functional utility of each type. Next we consider what categories of information are needed for echolocation, by addressing three chief pieces of information that bats need to extract from their echoes: 1) What is the nature of their target? 2) How far away are they from their target? and 3) Where is the target located? We then consider the acoustic cues that convey those categories of information.

In the second chapter we discuss the dominant organizational feature of the auditory system, the mapping of the sensory surface upon the principal nuclei of the auditory brainstem. We focus on the greatly expanded representation of a small frequency band in certain bats and explain why this region can be thought of as an "acoustic fovea". We show the similarities in tonotopic organization between certain species of bats and other mammals and illustrate how some of the major adaptations of bats, particularly the central representations of the so-called acoustic fovea, are variations of a basic mammalian arrangement.

The third chapter concerns the anatomy of the auditory systems of bats. We treat the auditory pathway in terms of two basic functions of the central auditory system, monoaural and binaural hearing. One of the fundamental principles which emerges from the study of the auditory system is that some pathways are designed to provide convergence of signals from the two ears, whereas others seem to maintain separation of the inputs from the two ears. We show how these pathways form systems which contain the structural basis for processing signals in the time and frequency domains. We describe the convergence of all the separate pathways from lower auditory regions at the inferior colliculus. Thus the inferior colliculus is a nexus in the auditory pathway where signals are integrated and reorganized before being transmitted to the thalamus and cortex. In the description of pathways to the inferior colliculus of the mustache bat we emphasize the greatly expanded 60 kHz isofrequency contour, the representation of the acoustic fovea in the midbrain. We then show why this adaptation is of particular value for elucidating the micro-organization of an isofrequency contour, the functional unit of each auditory nucleus.

The fourth chapter is a description of the response properties of auditory neurons. We consider the processes that generate the particular discharge characteristics of neuronal populations and the roles these have for coding target structure, target range and location. We discuss how the activity of ensembles of neurons could encode representations of target features and locations. Particular emphasis is given to the foveal regions of the auditory

5

system. We use the foveal region of the inferior colliculus as a model to illustrate how an isofrequency region is subdivided by groups of neurons with common response properties. We then show the significance of these features for creating a representation of target location.

The fifth chapter considers mechanisms activated during echolocation and how those mechanisms influence the neural processing of biosonar signals.

In the sixth chapter we bring together the features discussed in the previous sections and describe how the foveal regions illustrate the concept of a modular organization of the nervous system. Moreover, with specific illustrations we suggest how the monaural and binaural pathways could convey different attributes of the external world and what advantages accrue to such dichotomous systems for processing information. Finally, we consider some of the functional consequences of the massive convergence of projections from all of the lower brainstem nuclei at the inferior colliculus.

1.2 Historical Review

Bats are distinguished from all other mammals by their ability to fly, a distinction recognized by their classification into an order called Chiroptera, meaning "hand wing". It was not the flight of bats, however, that first attracted scientific attention, but rather their uncanny ability to avoid obstacles in the dark. As pointed out in the excellent historical reviews of echolocation by Robert Galambos (1942a) and Donald Griffin (1958), the scientific studies of bats begin with the question of orientation by the naturalists Lazzaro Spallanzani and Louis Jurine at the end of the eighteenth century. In a series of carefully controlled studies, they systematically eliminated each sensory system, except for the ear, as the means of orientation. They concluded that bats must orient through their sense of hearing. The conclusion of acoustic orientation, however, was not generally accepted by the scientific community. The French anatomist Cuvier championed the idea of a sixth sense, a form of touch arising in the skin from the movement of air, to explain the ability of blind bats to orient. This explanation dominated scientific thinking for more than a century. Indeed, the notion of acoustic orientation with sounds not detectable by the human ear seemed so outrageous as to elicit ridicule, a reaction which was most pointedly expressed by Montagu, who wrote in 1809: "To assent to the conclusions which Mr. Jurine has drawn from his experiments, that the ears of bats are more essential to their discovering objects than their eyes, requires more faith and less philosophic reasoning than can be expected of the zootomical philosopher, by whom it might fairly be asked, 'Since bats see with their ears, do they hear with their eyes?'" (quoted in Galambos 1942a).

In the early 1940's Donald Griffin and Robert Galambos repeated the earlier experiments of Spallanzani and Jurine and obtained the same results

(Griffin and Galambos 1941; Galambos and Griffin 1942). With the aid of new technologies, they also demonstrated that bats emit ultrasonic pulses in a characteristic fashion during obstacle avoidance. They observed that when approaching an obstacle the animals invariably increased the repetition of their pulses markedly while shortening the duration of their cries to prevent the pulse from overlapping with the echo. When very close to an object, the pulse repetition rate was often as high as 200 pulses/s. By monitoring cochlear microphonic potentials, Galambos (1942b) then showed that the bat's ear did, in fact, respond to the ultrasonic frequencies that are emitted for orientation. These findings settled the issue, and orientation accomplished through the use of biosonar was both demonstrated and universally accepted. Shortly thereafter, Griffin (1944) coined the term "echolocation" to describe this form of biological sonar.

1.3 The Orientation Calls of Bats

Microchiropteran bats compose the most diverse order of mammals on earth and are found in every region of the world except for the polar areas. Sixteen families and more than 800 species are recognized (Allen 1967). It is not surprising then, that a variety of biosonar signals have been documented from different species. Fortunately, almost all neurophysiological and neuroanatomical studies have been conducted on bats that employ one of two major types of biosonar signals: the FM signals and the long CF/FM signals that are described below.

Griffin and Galambos conducted their pioneering studies on little and big brown bats (Myotis lucifugus and Ephesicus fuscus), two closely related species common in the Northeastern United States. These bats orient by emitting loud but brief ultrasonic signals (Griffin 1958). The duration of each call is only 0.5 to 5.0 ms, and is emitted at a sound pressure level (SPL) of over 100 dB at about 10 cm from the bat's mouth. The calls have a characteristic frequency modulated (FM) structure, always sweeping downward about an octave throughout the duration of the signal. Significant energy is usually present in one or more harmonics. Representations of the big brown bat's orientation call are shown in Fig. 1.1 A. Such brief FM signals have been recorded from many species. Bats that employ this type of biosonar signal for orientation are referred to as "loud FM bats", or simply as "FM bats".

A second major type of orientation signal was discovered by the German zoologist Möhres (1953), who studied the echolocation system of the greater horseshoe bat. In contrast to the brief FM signals of the brown bats, horseshoe bats emit orientation calls that are typically 30–60 ms in duration and can be as long as 200 ms (Fig. 1.1 B). The major portion of the call is composed of a constant frequency (CF) portion at about 80 kHz, a pure tone in effect, and each call is terminated with a brief, 2 to 4 ms, FM portion that sweeps downward about 20 to 25 kHz. The calls are loud, at least 100 dB

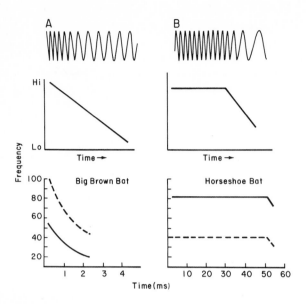

Fig. 1.1 A, B. Representations of orientation sounds emitted by loud FM bats (*A*) and long CF/FM bats (*B*). The *top panel* shows the fine structure of two artificial signals: a brief, downward sweeping frequency modulation signal and a long CF/FM signal. The *middle panel* shows time-frequency representations of these signals. Frequency is represented on the ordinate and time along the abscissa. The *lower panel* shows time-frequency representations of the orientation calls of the big brown bat (*left*) and the horseshoe bat (*right*). The *solid lines* show the frequencies containing the dominant energy in the signal. The *dotted lines* show the harmonics that have less energy. A time-frequency representation is commonly referred to as a sonogram or sound spectrograph

SPL, and have harmonics (e.g., see Neuweiler et al. 1980). Bats that utilize such biosonar signals are called "long CF/FM bats".

1.4 Nomenclature Used to Refer to Bats Throughout the Monograph

The majority of studies of the auditory systems of bats have been conducted on six species. In the spirit of addressing a wide audience, we shall refer, where possible, to each bat by its common name instead of using Linnaean nomenclature. Three of these species utilize loud FM cries for orientation, and the other three species are long CF/FM bats. The FM bats are: 1) the big brown bat, *Eptesicus fuscus* (family Vespertilionidae); 2) the Mexican free-tailed bat, *Tadarida brasiliensis* (family Molossidae); and 3) a bat for which there is no common name, *Molossus ater* (family Molossidae). The long CF/FM bats are: 1) the greater horseshoe bat, *Rhinolophus ferrumequinum* (family Rhinolophidae); 2) the rufous horseshoe bat, *Rhinolophus rouxi*

Table 1. Nomenclature used to refer to bats throughout the book

Common name	Genus-species	Type	Sonogram
Big brown bat	*Eptesicus fuscus*	loud FM	
Mexican free-tailed bat	*Tadarida brasiliensis*	loud FM	
no common name	*Molossus ater*	loud FM	
Greater horse-shoe bat	*Rhinolophus ferrumequinum*	long CF/FM	
Rufous horse-shoe bat	*Rhinolophus rouxi*	long CF/FM	
Mustache bat	*Pteronotus parnellii*	long CF/FM	

(family Rhinolophidae); and 3) the mustache bat, *Pteronotus parnellii* (family Mormoopidae). These bats are listed in Table 1, together with a sonogram of their echolocation calls.

1.5 How Bats Manipulate Their Signals During Echolocation

A universal modification, first described for FM bats by Griffin and Galambos, is an increase in the repetition rate of the cries as the bat detects

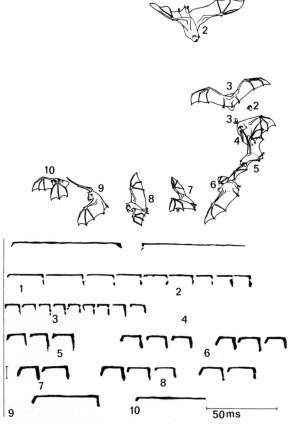

Fig. 1.2. Drawings from photographs of a horseshoe bat catching a moth. The *numerals* indicate sequential frames and thus indicate the position of the bat and the moth. Sonograms of the echolocation sounds emitted during flight are shown below. The bat and moth are shown in positions *2* and *3* (frame 1 is not shown). The moth is caught at position *4*. *Numbers below* echolocation sounds indicate position of the bat in the flight path. Note the long constant frequency component and the progressive shortening of this part of the orientation sound as the bat homes in on its target, shown in sonograms 1 through 3. The two very long orientation calls at the top (not numbered) were emitted before the bat detected the moth. (Neuweiler et al. 1980)

10

and pursues targets. Although the duration of the cries is also shortened during pursuit, the signals basically remain as brief FM pulses. How the auditory system responds to such signals is addressed in later chapters.

The long CF/FM bats, such as horseshoe and mustache bats, regulate both the repetition rate and duration of their pulses (Fig. 1.2). When a long CF/FM bat detects an object, it increases the repetition rate of its cries. As it continues to home in on its target, it progressively shortens its orientation calls, although the bat never completely eliminates the CF component (Novick and Vaisnys 1964; Neuweiler et al. 1980). Thus the duration is never as short nor the repetition rates as high as those achieved by the FM bats.

The most remarkable modification is that long CF/FM bats adjust the frequency of the CF component of their pulses to compensate for Doppler shifts of the echoes that reach their ears (Fig. 1.3). This behavior, called Doppler-shift compensation, was discovered by Hans-Ulrich Schnitzler (1967) in horseshoe bats and is the expression of an extreme sensitivity for motion. The motion is signaled by the difference in frequency between the emitted and echo components of the CF signals. The CF emitted by a bat that is not flying is almost constant, varying by only 50–100 Hz from pulse to pulse (Schuller et al. 1974; Henson et al. 1980; Henson et al. 1982). During flight, echoes reflected from stationary objects in the environment are shifted upward in frequency due to the relative movement of the bat towards the stationary background. Both horseshoe and mustache bats (Schnitzler 1970) compensate for the Doppler shifts in the echoes by lowering the frequency of pulses emitted subsequently by an amount nearly equal to the upward fre-

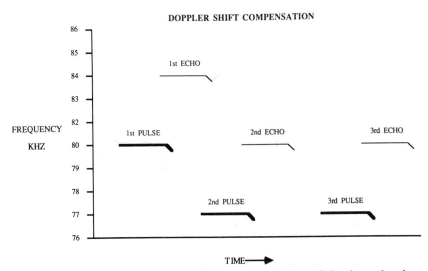

Fig. 1.3. Schematic illustration of Doppler-shift compensation in a flying horseshoe bat. The first pulse is emitted at about 80 kHz, and the Doppler-shifted echo returns with a higher frequency, at about 84 kHz. The horseshoe bat detects the difference between the pulse and echo and lowers the frequency of its subsequent emitted pulses by an amount almost equal to the Doppler shift. Thus the frequencies of the subsequent echoes are held constant and return at a frequency very close to that of the first emitted pulse

11

quency shift in the echo. Consequently, the bat clamps the echo CF component and holds it within a narrow frequency band that varies only slightly from pulse to pulse. It is remarkable that horseshoe bats can detect, and compensate for, pulse-echo frequency differences as small as 50 Hz (Simmons 1974; Schuller et al. 1974). These bats then, can discriminate a frequency change of only 0.06% and therefore have a frequency acuity about an order of magnitude better than most other animals.

Bats also regulate the intensity of the echoes which reach their ears by adjusting the intensity of their emitted pulses. Kobler and his colleagues (Kobler et al. 1985) showed that when mustache bats are swung on a pendulum, they emit orientation calls as they move towards a fixed target. As the bat approaches the target it systematically decreases the intensity of its orientation call. Apparently the bat adjusts the amplitude of its pulse so that the echo that returns is of constant intensity. This response resembles Doppler-shift compensation in that the bat seems to optimize a feature of the echo to enhance signal analysis.

1.6 Echolocation Signals Are Tailored to the Habitat in Which the Bat Hunts

Studies by Gerhard Neuweiler and his colleagues (Neuweiler 1983, 1984b) have shown that the use of a particular form of biosonar signal is closely correlated with the habitat in which a species normally hunts. Loud FM bats always hunt in the open sky, where they capture flying insects by scooping them out of the air with their wings or interfemoral membranes, the thin layer of skin stretched between their legs (Webster and Griffin 1962). Presumably the high intensity of their emitted signals increases the range at which they can track their prey.

The advantage of using a Doppler-based sonar system is that it enhances the ability to hunt in acoustically cluttered environments by maximizing the bat's ability to distinguish a fluttering insect from background objects and by providing an effective means for recognizing particular insect species. All long CF/FM bats that have been studied hunt for flying insects in dense foliage beneath the forest canopy (Bateman and Vaughan 1974; Neuweiler 1973, 1984). The behavior of Dopplershift compensation indicates that the auditory systems of long CF/FM bats are sensitive to the frequency changes induced by motion. However, flight speed differences between the bat and its prey do not represent the total extent of movements a hunting bat would encounter. Specifically, flying insects beat their wings to stay aloft. The motion of the wings creates periodic Doppler shifts, and thus periodic frequency modulations are superimposed on the CF component of the echo (Fig. 1.4). In addition, the wing motion, or flutter, presents a reflective surface area of alternating size, thereby also creating periodic amplitude modulations, or

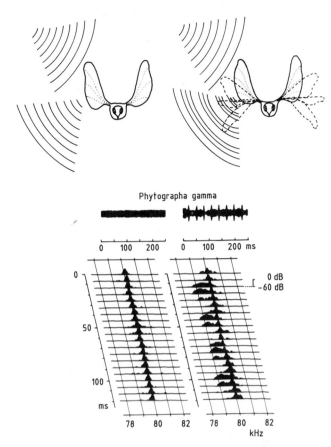

Phytographa gamma

Fig. 1.4. Upper panel Drawings to illustrate the echoes reflected from an insect in which the wings are stationary (*left*) and from an insect whose wings are in motion (*right*). *Lower panel* Oscillograms (*above*) and real time spectra (*below*) of echoes from a nonflying (*left panel*) and flying (*right panel*) noctuid moth, *Phytographa gamma*. The insect was tethered an oriented at an angle of 30 degrees in the acoustical beam of a loudspeaker transmitting a continuous 80-kHz tone. Note that the echo returning from the nonflying moth was simply an unmodulated tone of 80 kHz, whereas the echo reflected from a moth while its wings were beating had pronounced amplitude and frequency modulations. The amplitude modulations, or glints, are apparent in the oscillograms of the echoes from the flying moth. (Data from Schnitzler and Oswald 1983)

"glints", in the echo CF component. There is substantial evidence, reviewed below, that the periodic frequency and amplitude changes inform the bat that its target is indeed an insect, and that the profile of the frequency and amplitude modulations may provide important information for discriminating among different insect types.

The long CF/FM bats "see" much of their world through the narrow window of a small band of frequencies, around 60 kHz for mustache bats and 80 kHz for horseshoe bats. Not surprisingly, many of the pronounced specializations in the auditory systems of long CF/FM bats are regions designed for processing the CF component of the echo (e.g., Suga 1978; Schnitzler and Oswald 1983; Neuweiler 1984; Pollak et al. 1986). These adaptations were first seen in audiograms of evoked potentials recorded from the inferior colliculus of mustache bats by Alan Grinnell (1967) and of horseshoe bats by Gerhard Neuweiler (1970). These pioneering reports inspired subsequent studies of the auditory pathway in these bats. In the following chapters we discuss how the adaptations revealed by these subsequent studies illustrate basic features of anatomical organization and mechanisms for information processing.

1.7 What Information Do Bats Extract
from their Biosonar Signals?

The biosonar systems of bats are exceptionally sophisticated, permitting these animals not only to orient in space but also to recognize and dis-

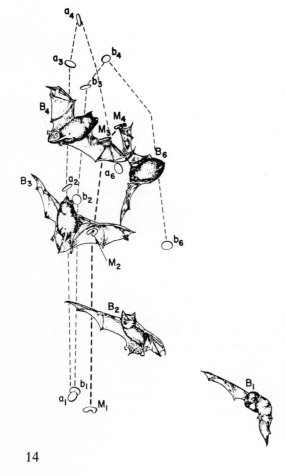

Fig. 1.5. Little brown bat (B) selecting a mealworm (M) out of a clutter of two discs (a and b). The positions of the bat and objects are indicated by subscripts. For example, B_1 refers to the bat's position relative to targets a_1, b_1 and M_1. (Data adapted from Webster and Durlach 1963. Figure from Schnitzler and Henson 1980)

14

criminate among multiple targets with remarkable precision and speed. These abilities are illustrated by experiments in which little brown bats were trained to catch mealworms tossed into the air (Webster and Durlach 1963; Griffin 1967). When several plastic discs, each with dimensions similar to a mealworm, were presented together with a mealworm, the bat learned to catch the mealworm and ignore the discs (Fig. 1.5). No doubt the bats use biosonar to "see" the size, shape and texture of objects, as well as to make accurate assessments of the spatial location, range, velocity and trajectory of its target. Below, we briefly discuss the cues bats use to assess these features.

1.8 Cues Bats Use to Extract Information from Biosonar Signals

1.8.1 Recognition of Targets

The cue FM bats use for target recognition is the spectral composition of the echo. The echoes from different objects have characteristic spectra created by the different reflective and absorptive properties, as well as the shape and texture of the target. The mealworms that Griffin (1967) trained his bats to catch, for example, generated echoes having a different spectrum from those generated by the plastic discs that were presented simultaneously. Subsequent behavioral experiments by Simmons et al. (1974) showed that bats can make extremely fine discriminations based on the spectral composition of the echoes (Fig. 1.6). They trained big brown bats to discriminate a plate having holes of 8 mm deep from other plates with hole depths between 6.5 and 8.0 mm. The echoes from each of these plates differed in the spectral distribution of energy but not in overall intensity. The interference between the sounds reflected from the surface and the interior of the hole in each target produced a characteristic echo spectrum that was related to hole depth (Fig. 1.6). The bats detected differences in hole depth smaller than 1 mm.

The results of Simmons and his colleagues were recently confirmed by Sabine Schmidt (1988), who used electronic methods to generate phantom echoes from two loudspeakers. Each bat was trained to direct its calls toward one of two loudspeakers from which the phantom echoes were presented. The bat was rewarded for choosing one of the phantom targets. The phantom targets were produced by adding the orientation call to a time delayed version of itself. The internal delay between the two replicas of the orientation sounds was varied in steps of 81 ns. This electronic manipulation simulated the interference patterns produced by reflections from two planes set at different distances, conditions analogous to those of the experiment of Simmons et al. (1974) described above. The phantom target was then played back to the bat with an absolute delay of 4.0 ms, so that phantom target appeared at an absolute distance of 1.34 m from the bat. The bat's task was to differentiate phantom targets having different internal delays from a phan-

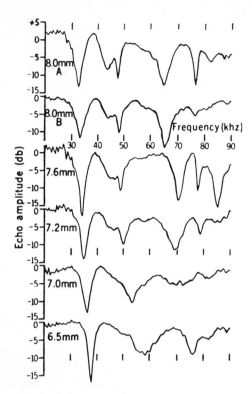

Fig. 1.6. Echo spectra from six different targets. Each target was a flat Plexiglas plate with holes drilled at different depths. Big brown bats were trained to discriminate a plate with holes of 8-mm depth from other plates with hole depths between 6.5 and 8.0 mm. The bats were able to distinguish between the plate with 8.00-mm holes and other plates as long as the depths differed by 0.6 mm or more. The echo spectra from each of the targets is shown. Amplitudes of the target spectra are expressed in dB relative to those of a smooth target. Note that the spacing of the amplitude peaks varies with hole depth. (Simmons et al. 1974)

tom target having an internal delay of 7.77 μs. Schmidt found that bats can discriminate between two targets in which the internal delays differ by only 1 μs. This difference corresponds to two echoes reflected from surfaces differing in depth by only 0.2 mm. Of particular importance is that the internal delays that the bats could discriminate also produced signals having notches in their power spectra that differed by several kHz. These results support the hypothesis that bats use the spectral composition of echoes to distinguish among flying insect prey and for identifying other objects on the basis of "surface texture" (Neuweiler 1984).

The long CF/FM bats probably also use spectral notches in the FM component to recognize surface texture of solid objects, but in addition, they use the patterns of amplitude and frequency modulations in the CF component of their echoes for prey recognition (Fig. 1.7). The hypothesis that the CF component conveys information for prey recognition was originally proposed by Hans-Ulrich Schnitzler (1970) and Gerd Schuller (1972). Over

16

Fig. 1.7. Drawing of the emitted pulse of a mustache bat and the frequency modulations in the echo due to the wing movements of a nearby moth. The emitted constant frequency component is depicted as a series of regularly spaced waves. The echo reflected from the beating wings of the moth is shown as a series of irregularly spaced waves that are repeated periodically (i.e., the frequency modulations)

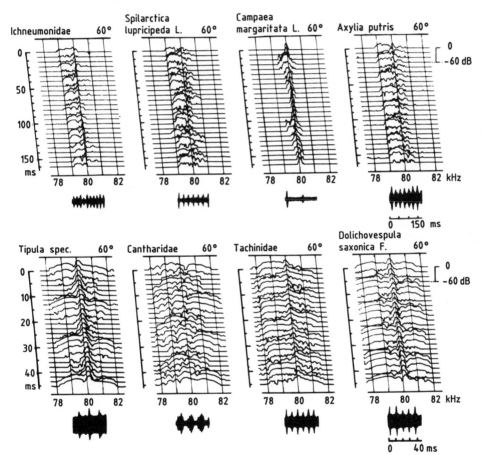

Fig. 1.8. Real-time spectra and oscillographs to show amplitude modulations of echoes produced by the wingbeat motion of eight different insect species. Each insect was flying while tethered and oriented at the same angle relative to the acoustical beam from a loudspeaker transmitting a continuous 80-kHz tone. Notice that the patterns of both amplitude modulations in the oscillographs and real time spectra differ noticeably among the insect species. (Schnitzler et al. 1983)

17

the past decade, Schnitzler and his colleagues have conducted a wide range of behavioral studies concerned with flutter detection in long CF/FM bats. These elegant and comprehensive studies provide three major lines of evidence that prey recognition is indeed the primary purpose of the CF component.

First, information about the species of insect is coded in the periodic modulation patterns. Schnitzler and his colleagues (Schnitzler et al. 1983) showed this by beaming long tones onto different species of fluttering insects and recording the echoes (Fig. 1.8). They found that echoes from each species have an "acoustic signature" conveyed as a characteristic power spectrum and temporal pattern of amplitude peaks.

Second, long CF/FM bats ignore stationary insects whose wings are not beating or insects whose wingbeats are too slow. However, the bats readily pursue insects with more rapid wingbeats (Airapertjantz and Konstantinov 1974; Goldman and Henson 1977; Link 1986). In addition, these bats are selective in their choice of fluttering insects, preferring certain insect species over others.

Third, long CF/FM bats are extremely sensitive to fluttering targets. Schnitzler and Flieger (1983) showed that horseshoe bats can discriminate an 83 kHz constant frequency signal from a signal in which the frequency was sinusoidally modulated by only ± 12 Hertz around an 83 kHz carrier! The insects that horseshoe bats prey upon can create modulation depths of ± 1000 Hz or more (for example see Figs. 1.4 and 1.8), which are far above the bat's detection threshold. All of this supports the idea that long CF/FM bats detect and recognize insect species from the modulation patterns created by the motion of their fluttering wings.

1.8.2 Localization of Targets in Space

1.8.2.1 Target Range

In an elegant series of studies, Simmons (1971 a, b, 1973) showed that range is determined through evaluations of the temporal intervals between the pulse and echo FM components. His original experiments utilized the big brown bat, *Eptesicus fuscus,* but in subsequent experiments he showed that long CF/FM bats also use the temporal interval between the pulse and echo of their terminal FM component for range assessments (Simmons 1971 b).

Actually, Simmons determined the bat's ability to discriminate the difference in the range between two targets. In his experiments, the bat was required to scan two identical targets with its orientation calls and choose the closer of the two (Fig. 1.9). The calls were monitored by a microphone placed close to the bat's mouth, and phantom echoes were then played back to the bat from loudspeakers situated at a constant distance from the bat. The echoes were identical in all respects but one: the temporal interval between the pulse and echo, which was varied electronically. When the pulse-echo interval from one target was long and the interval from the other substantially

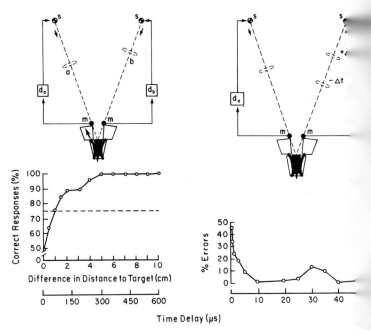

Fig. 1.9. Experimental arrangements used by Simmons for determining sensitivity for pulse echo delays. *Left* Two-choice apparatus for presenting electronically simulated targets where the phantom echoes had different delays. Sonar pulses are detected by microphone (*m*), and are broadcast back to the bat from loudspeakers (*s*) after a short delay determined by delay lines d_a and d_b. Delay d_a is shorter than d_b, so that phantom target *a* appears closer than the phantom target *b*. *Below* is shown the bat's performance at different delays, which are also shown as differences in target distance. (Data from Simmons 1973). *Right* Arrangement for echo "jitter" experiment (Simmons 1979). Only the channel toward which the bat directed its sonar pulses produced a phantom echo. In this example, "echoes" from successive sonar pulses directed at channel *b* had delay times that alternated between $+ \Delta t$ and $- \Delta t$. "Echoes" from pulses directed at channel *a* had a fixed time delay. *Below* is shown performance of bat in percent errors as a function of the difference between $+ \Delta t$ and $- \Delta t$. (After Suthers and Wenstrup 1987)

shorter, the bat chose the correct, or closer, target in 90–95% of the cases. As the interval between the pulse and echo was reduced, performance deteriorated progressively. When the difference between the pulse echo intervals became less than about 75 μs, corresponding to a difference in target distance of slightly less than 1 cm, performance fell below the criterion level of 75% correct choices.

In a more recent experiment Simmons (1979) again employed the phantom echoes, but in a somewhat different paradigm and showed that bats have a much greater acuity for discriminating pulse echo intervals than was previously believed (Fig. 1.9). In this experiment the bats were again required to scan two phantom targets and the task was to choose the "stationary target". The echoes from one of the phantom targets returned at a constant delay (Δt), as they would from a stationary target. In contrast, the echo delays from the other phantom target alternated, where one echo returned at

ay $(+\Delta t)$, and the next echo returned at a slightly shorter
he target appeared to "jitter", and the task of the bat was
een the stationary and "jittering" targets. The magnitude
ranged from ± 40 to 0 μs. Simmons showed that the bats
choes having delay differences of 1 μs from echoes that
tant delay. We can conclude from these studies that bats
interval to assess target range, and they do so with
cy

imuth

e assessment of a sound's localization in azimuth and eleva-
hed through binaural processing of interaural disparities.
ration acuity in the horizontal plane, or azimuth, suggest
criminate the angular difference of targets separated by as
grees (Peff and Simmons 1971; Simmons et al. 1983). The in-
ies are generated by the physical properties of the animal's
Two chief types of interaural disparities are available to
arities in sound pressure level which are usually called inter-
ifferences (IIDs); and 2) disparities in arrival time (Fig. 1.10).
siderable controversy about which cues provide bats with in-
t target azimuth. Unlike the ranging experiments, where all
the temporal interval between pulse and echo were held con-
ve been no behavioral experiments conducted on bats to
ability to localize the azimuth of a sound source on the basis
ferences in time or intensity. The experiments in which these
re shown to be the primary cues for localizing a sound in the

ings to illustrate generation of interaural time disparities and interaural in-
ies. At *left* sound waves reach the bat's ears from a source directly ahead. In
unds reaching both ears will be of equal intensity and will arrive at the same
path lengths, indicated by the *broken lines* from the source to the ears, are
und source is displaced to one side, as shown on the *right,* the sound waves
r ear unimpeded, but the head and ears block most of sound to the farther
astic shadow makes the sound more intense in the closer ear than in the
d thus creates an interaural intensity disparity. In addition, the sound path
ar is shorter than the path to the farther ear thereby creating a difference in
es of the sound at the two ears, indicated by the *different lengths of the broken*

horizontal plane were those in which human observers listened to signals via earphones (see Mills 1972 and Hafter 1984 for excellent reviews). Understandably, such experiments have not yet been repeated in bats, and the effects of variations in one of these cues while holding the other constant have not been determined. In the absence of such information, our analysis will rely on assessing the disparities that are available to bats and then consider whether their nervous systems are able to process the interaural disparities that the bats would normally experience.

Interaural Intensity Disparities. The short wavelengths of ultrasonic frequencies yield disparities in sound pressure at the two ears when sounds originate from positions off the midline (Grinnell and Grinnell 1965; Henson 1967; Fuzessery and Pollak 1984, 1985; Harnischfeger et al. 1985). The animal's head is an obstacle that creates a substantial acoustic shadow, thereby reducing the sound pressure at the ear farther from the sound source (Fig. 1.10). In addition, the elaborate ears of bats are highly directional and amplify sounds emanating from certain positions in space in a frequency-dependent manner. These features combine to create interaural intensity disparities that can be as great as 30 to 40 dB and constitute the most pronounced cue available to bats to localize targets.

Interaural Arrival Time Disparities. The difference in the distance that sound must travel to reach each ear produces disparities in the arrival time of the sound at the two ears. These disparities therefore vary as a function of the azimuth of the sound source. It is convenient to think of two types of time disparities. With the first type, called onset disparity, the important cue is the temporal difference between the leading edges, or onsets of the signals. The time difference of the signal onset at the two ears is encoded and compared by the binaural portion of the auditory system. Because the heads of all bats are small, the range of onset disparities produced by sounds originating from different azimuthal positions are also quite small. The headwidth of the bats considered in this monograph yield onset disparities of 50 μs, at most. For sounds that originate within ±10 degrees from the midline, the onset disparities would be only about 5 μs. The ability of the bat's auditory system to process these temporal disparities is considered in later sections.

The second form of temporal disparity, called the ongoing time disparity or simply phase disparity, is based on the periodicity of a longer repeating waveform. The waveform is encoded at each ear by neurons that respond only to the positive phase of the waveform. The task of the nervous system is to determine whether or not similar points (i.e., phase) are coincident at the two ears. Sounds originating along the midline will arrive at the two ears at exactly the same time, and thus the phase of the signals at the two ears will be aligned perfectly. Sounds situated at any other position will generate signals that are out of phase at the two ears. In order to assess disparities in ongoing time, or phase, the nervous system must code the phase of the waveform and compare the temporal sequence of these phase-locked discharges as they converge from the two ears. However, binaural phase comparisons provide un-

21

ambiguous directional information only if the interaural distance is less than one-half of the wavelength. High frequencies have short periods, creating phase disparities that can have the same value at more than one azimuthal location, unless headwidth is sufficiently small. For bats having a headwidth of 1.0 cm, the theoretical upper limit for phase comparisons is about 30 kHz. The fact that phase-locking is maintained only up to about 4–5 kHz in the cat (Kiang et al. 1965), for example, sets an even lower limit for phase comparisons. We do not know the upper limit for phase-locking in bats, but it seems unlikely to be higher than 5 kHz.

The ultrasonic world of bats provides additional signals suitable for processing ongoing time disparities. These are the frequency and amplitude modulations of the echo CF components of the long CF/FM bats. Interaural comparisons of ongoing time disparities for these signals require that the nervous system code for the phase of the modulating waveform, rather than phase of the carrier frequency (the signal's fine structure). The wingbeat frequencies of the insects these bats prey upon vary from about 35 Hz to a few hundred Hertz and thus generate modulation waveforms having wavelengths much longer than the headwidths of the bats. However, the headwidth of a mustache or horseshoe bat will produce an interaural time disparity of 25–35 μs, at most. Due to these very small time differences, the bat will have to detect interaural phase differences of only a few μs from signals having periods of 10–20 ms or more. The use of modulated signals for localizing targets has only recently received attention in neurophysiological studies.

1.8.2.3 Echolocation Maximizes Interaural Disparities

In general, the interaural cues for assessing location are produced by the interaction of sound waves and the physical properties of an animal's head and ears. However, the use of an active sonar system allows bats to limit and also manipulate the acoustic cues they receive. The range of potential interaural disparities is limited by the emission of orientation pulses as a more or less narrow beam of sound which helps to focus the bat's acoustic field upon objects within a small region of space directly in front of its flight path (Griffin 1958; Simmons 1969; Griffin and Hollander 1973; Shimozawa et al. 1974; Schnitzler and Grinnell 1977; Grinnell and Schnitzler 1977). The echoes returning from these fairly restricted regions of space interact with highly directional ears to enhance the intensity disparities that bats receive. Some bats, most notably horseshoe bats, also move their ears in an alternating fashion with the emission of each echolocation call (Griffin et al. 1962; Pye et al. 1962). These movements alter the intensity differences at the two ears in a complex fashion (Neuweiler 1970). The range of ear movements in other bats, such as the mustache bat, is much less pronounced (Schnitzler 1970), and yet in other bats, such as molossids, ear movements are minimal or not present at all. In Chapter 4 we consider the processing of interaural disparities in *M. ater* and mustache bats. Since the role of ear movements for localization is not well understood in these species, we only consider the possible intensity disparities that would be produced without ear movements and the coding of these disparities in the auditory pathway.

1.9 Biosonar Signals and the Phylogeny of Bats

The six species of bats upon which the majority of neural studies have been conducted belong to four different families. Earlier we emphasized that bats which emit a particular type of echolocation call manipulate their signals in a similar fashion and use the same cues to extract similar types of information from their echoes. For the six species considered here the general rule is that closely related bats emit similar echolocation calls and have similar auditory systems. For example, the two species belonging to the family Rhinolophidae, the rufous horseshoe bat and greater horseshoe bat, emit similar biosonar signals and have auditory systems that are nearly indistinguishable. For these reasons the two species are usually referred to interchangeably as "horseshoe bats".

In contrast, some species that are distantly related employ the same type of biosonar signals. These species usually have markedly different specializations for processing similar aspects of their biosonar signals. This point is illustrated by the two major groups of long CF/FM bats, mustache and horseshoe bats. Both species exploit the same environmental niche: they hunt flying insects in acoustically cluttered places and they utilize a Doppler-based sonar system for this purpose. However, their behavioral similarities are, in some cases, due to quite different structural features. For example, the cochleae of mustache and horseshoe bats are quite different. The differences in the auditory systems are a consequence of the independent evolution of the two groups; mustache bats evolved in the Neotropics while horseshoe bats evolved in Europe and Asia. Our point is that, while the biosonar calls are very similar in the two groups, assumptions about structural and functional homologies can be misleading unless appropriate criteria are satisfied with anatomical and physiological data.

Chapter 2

Tonotopic Organization

2.1 Introduction

Here we begin a discussion of the bat's auditory system based upon three chief organizational features. The first is that the dominant feature of the system is a tonotopic organization, a remapping of the cochlear surface upon most of the nuclear groups of the primary auditory pathway. The second is that the principal pathways originate from the three major divisions of the cochlear nucleus: the anteroventral cochlear nucleus (AVCN), the postero-ventral cochlear nucleus (PVCN) and the dorsal cochlear nucleus (DCN) (Fig. 2.1). Each pathway ascends in parallel with the others and has a unique pattern of connectivity with the subdivisions of the superior olivary complex and the nuclei of the lateral lemniscus. All of these pathways ultimately terminate in an orderly fashion in the central nucleus of the inferior colliculus, where the multiple tonotopic maps of the lower nuclei are reconstituted into a single tonotopic arrangement. The third feature is that the projections emanating from the AVCN form both monaural and binaural pathways, and these two pathways convey information about different aspects of the external world. The latter two organizational features will be treated in more detail in the next chapter. In this chapter we describe the tonotopic arrangement in the major auditory nuclei of the bat's brainstem, beginning with a consideration of the frequency representation in the cochlea and auditory nerve. The specialized tonotopy of the long CF/FM bats is given primary consideration and is contrasted with the more general tonotopic organization of FM bats. For purposes of orientation, a photograph of the mustache bat's brain is shown in Fig. 2.1, together with a schematic drawing of the main auditory brainstem nuclei.

2.2 The Cochlea and Auditory Nerve

The initial action of the cochlea is a conversion of frequency-to-place, whereby the power spectrum of an orientation sound is transformed into a spatial pattern of activity along the cochlear partition. The cochleae of long CF/FM bats, however, do not give equal representation to each frequency. The length of cochlear partition devoted to the dominant frequency of the CF component, 60 kHz in mustache bats and 80 kHz in horseshoe bats, is ex-

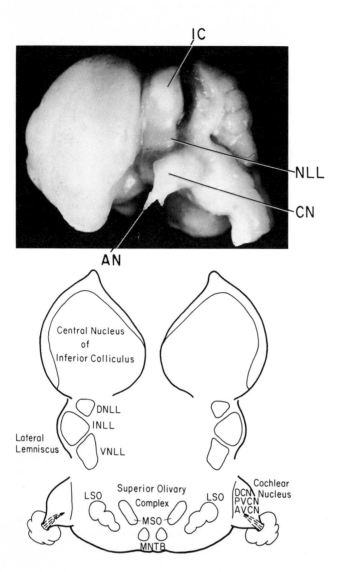

Fig. 2.1. Upper panel Photograph of the brain of the mustache bat with the cerebellum partially removed to allow visualization of the auditory structures. The left side of the brain is shown with the cerebrum at the left and the brainstem at the right. The auditory nerve (*AN*), cochlear nucleus (*CN*), nuclei of the lateral lemniscus (*NLL*) and inferior colliculus (*IC*) are labeled. *Lower panel* Drawing of a transverse section of the mustache bat's brainstem to illustrate the location of the principal auditory nuclei: anteroventral cochlear nucleus (*AVCN*), dorsal cochlear nucleus (*DCN*), dorsal nucleus of the lateral lemniscus (*DNLL*), intermediate nucleus of the lateral lemniscus (*INLL*), lateral superior olive (*LSO*), medial nucleus of the trapezoid body (*MNTB*), medial superior olive (*MSO*), posteroventral cochlear nucleus (*PVCN*) and ventral nucleus of the lateral lemniscus (*VNLL*)

26

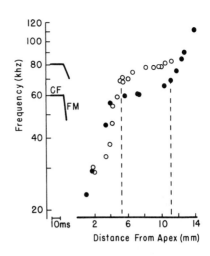

Fig. 2.2. Comparison of frequency-place maps of the cochlea of the mustache bat (*filled circles*) and the horseshoe bat (*open circles*). Distance along the basilar membrane, measured from the cochlear apex, is plotted on the abscissa and the frequency represented at each place is on the ordinate. Both maps exhibit a steep slope in the low frequency range followed by a long stretch of basilar membrane devoted to a narrow frequency band, around 60 kHz for the mustache bat and around 80 kHz for the horseshoe bat (bracketed by *broken lines*). Sonograms of the dominant component of each bat's orientation call are shown on the *left*. (After Kössl and Vater 1985)

panded and has numerous morphological and physiological specializations (Bruns 1976a, b; Henson 1978; Bruns and Schmieszek 1980; Wilson and Bruns 1983; Kössl and Vater 1985; Leake and Zook 1985; Vater et al. 1985; Vater 1987; Zook and Leake 1988) (Fig. 2.2). These features, coupled with behavioral specialization of Doppler-shift compensation, whereby the bat ensures itself that the echo frequencies will coincide with the expanded cochlear region, suggest that the elongated segment of the cochlear partition functions as an "acoustic fovea" (Schuller and Pollak 1979; Bruns and Schmieszek 1980). The mechanisms for producing the acoustic foveae in the two species are, however, considerably different from one another.

The distinctive features of the horseshoe bat's cochlea are anatomical. One of the most prominent features is an abrupt change in the thickness and width of the basilar membrane in the basal turn, at a locus close to the place where 80 kHz is represented (Bruns 1976a, b; Vater 1987) (Fig. 2.3). These features undoubtedly produce sharp changes in membrane stiffness in this region and thus must have a substantial influence on the mechanical action of the basilar membrane (Wilson and Bruns 1983; Vater 1987). Changes in basilar membrane dimensions are far less pronounced in the mustache bat's cochlea (Henson 1978; Kössl and Vater 1985; Vater 1987) (Fig. 2.3).

Physiological responses, in contrast, are far more pronounced in the cochleae of mustache bats than in horseshoe bats. These features are reflected in cochlear microphonic potentials (Pollak et al. 1972, 1979; Suga et al. 1975; Suga and Jen 1977; Schnitzler et al. 1976; Henson et al. 1985), the

27

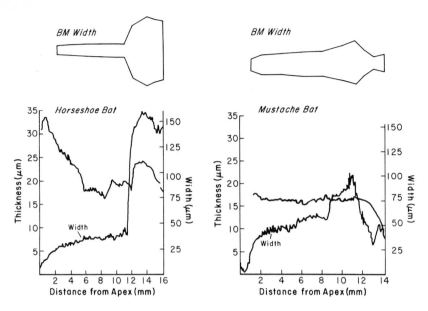

Fig. 2.3. Comparison of morphological specializations in the cochleae of horseshoe (*left*) and mustache bats (*right*). In *lower panels,* the width and thickness of the basilar membrane (*BM*) along the length of the cochlear partition are shown graphically. The way the basilar membranes would appear if they were uncoiled and if one could look down upon them are shown in the *upper panels.* Note the pronounced changes in width along the horseshoe bat's basilar membrane and the less pronounced changes in width in the mustache bat's cochlea. Data for horseshoe bat from Bruns (1976a), data for BM width in the mustache bat from Henson (1978), BM thickness data from Kössl and Vater (1985). (After Vater 1987)

summed responses of the outer hair cells of the organ of Corti (Dallos et al. 1972). The cochlear microphonic potentials evoked by tone bursts from the mustache bat's cochlea display a pronounced resonance at 60 kHz and are especially large when evoked by a narrow range of frequencies around 60 kHz (Fig. 2.4). The resonance is also manifest in the sharp tuning at 60 kHz of the mustache bat's cochlear microphonic audiogram (Pollak et al. 1972; Fig. 2.5). This observation is noteworthy because the resonant frequency and the frequency at which mustache bats hold their echo CF component when compensating for Doppler shifts are nearly identical (Henson et al. 1980; 1982). In horseshoe bats, the cochlear microphonic potentials evoked by tone bursts display no such resonance (Henson et al. 1985). Unlike the mustache bat, there is no prominent resonance at 80 kHz, and the tuning of the cochlear microphonic audiogram is not nearly as prominent as it is in the mustache bat (Fig. 2.5) (Schnitzler et al. 1976). Although the mechanisms of the specializations of the acoustic foveae in the horseshoe and mustache bats are not well understood, the end result is the same in both species: a sharp frequency selectivity in the neurons which innervate the hair cells in the cochlear regions devoted to the CF component of their orientation calls.

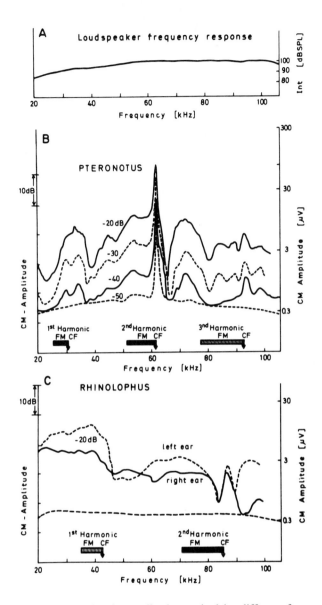

Fig. 2.4 A–C. Comparison of cochlear microphonic amplitudes evoked by different frequencies for mustache bat (*B*) and horsehoe bat (*C*). The stimulus intensities, shown on the *left* of each cochlear microphonic graph, are given in dB below the maximum output of the loudspeaker, shown in the *top panel* (*A*). Note the sharply tuned absolute amplitude magnitude around 60 kHz for the mustache bat, and the small peak around 83 kHz in the horseshoe bat. Note also the absolute differences in amplitude of cochlear microphonic potentials. *Dotted line* at bottom of C gives noise level of equipment. (Henson et al. 1985)

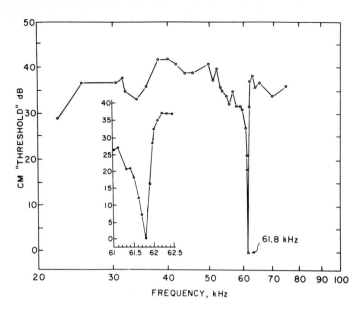

Fig. 2.5. Cochlear microphonic audiogram of mustache bat. Each *point* represents the "threshold" response of cochlear microphonic potential for that frequency. The very sharp tuning of the audiogram is apparent. *Insert* shows expanded view of threshold changes around 61.8 kHz. (Pollak et al. 1972)

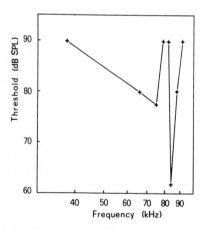

Fig. 2.6. Tuning curve of a filter neuron from the auditory nerve of a horseshoe bat. (Neuweiler and Vater 1977)

The response properties of auditory nerve fibers in the mustache and horseshoe bat have been the subject of several reports (Suga et al. 1975, 1976; Suga et al. 1975; Suga and Jen 1977; Suga and Manabe 1982). These studies suggest that the elegance of the cochlear frequency-to-place transform is preserved in the population of auditory nerve fibers. The acoustic fovea in each species is represented in the auditory nerve as a disproportionately large population of fibers that are highly selective for 60 kHz in mustache bats and

30

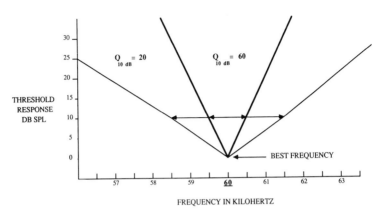

$$Q_{10\,dB} = \frac{\text{BEST FREQUENCY}}{\text{TUNING CURVE AT } 10\ dB\ \text{ABOVE BF THRESHOLD}}$$

Fig. 2.7. Diagram to show how $Q_{10\,dB}$ values are calculated from two hypothetical tuning curves

for 80 kHz in horseshoe bats (Fig. 2.8). The frequency selectivity of an auditory neuron is seen in its tuning curve, the graph showing the neuron's discharge threshold for different frequencies (Fig. 2.6). Each tuning curve can be described by a quality factor, or $Q_{10\,dB}$ value, which reflects its sharpness or frequency selectivity, and by its best frequency, the frequency to which the neuron is most sensitive (Fig. 2.7). The $Q_{10\,dB}$ value is calculated by dividing the best frequency by the frequency bandwidth of the tuning curve at 10 dB above threshold. Thus the higher the $Q_{10\,dB}$ value, the sharper the tuning. The tuning curves of neurons having best frequencies at 60 kHz in mustache bats and at 80 kHz in horseshoe bats are extremely sharp, with $Q_{10\,dB}$ values as high as 300–400 (Fig. 2.8). Because of the limited range of frequencies to which these neurons respond, they have been called "filter neurons" (Neuweiler and Vater 1977), a term that will be used here. Units having best frequencies above of below the dominant constant frequency component of the orientation cry commonly have small $Q_{10\,dB}$ values that range from 3–18. The complete spectrum of frequencies to which the cochlea responds is, of course, re-represented in the auditory nerve, but the representation is greatly weighted in favor of the filter units.

The peripheral auditoy systems of FM bats, such as the brown bats, show few, if any, of the unique features so characteristic of long CF/FM bats. Anatomical studies have revealed no apparent specializations of the cochlea (Ramprashad et al. 1978), at least not of the sort seen in the horseshoe bat. Consistent with these observations, the $Q_{10\,dB}$ values of peripheral fibers in FM bats are comparable to those of nonecholocating mammals and range from about 3–18, with very few above 20 (Suga 1973) (Fig. 2.8).

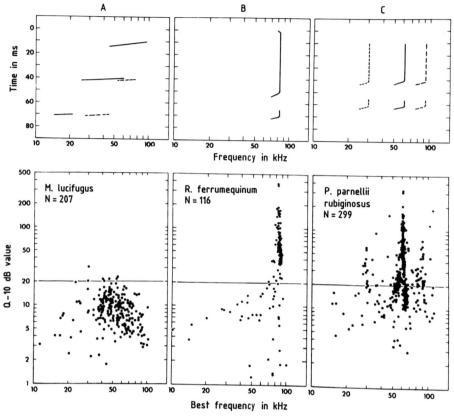

Fig. 2.8 A–C. Sonograms (*upper figures*) and distributions of Q_{10dB} values of neurons (*lower graphs*) in the auditory nerve of three bats: *A* the little brown bat; *B* the horseshoe bat; and *C* the mustache bat. Note the horseshoe bat (*B*) has a disproportionately large number of neurons having high Q_{10dB} values at around 80 kHz, the frequency of its CF component. The mustache bat (*C*) also has a pronounced over-representation of neurons having large Q_{10dB} values at 60 kHz, a frequency corresponding to the dominant CF component of its species. A smaller proportion of neurons at 30 and 90 kHz also have high Q_{10dB} values. (Suga and Jen 1977)

2.3 Cochlear Nucleus

As it exits the cochlea, the auditory nerve is a single pathway, but as it enters the cochlear nucleus, its fibers bifurcate to form ascending and descending branches which innervate the three major divisions of the cochlear nucleus, the AVCN, the PVCN and the DCN (Harrison and Feldman 1970; Lorente de Nó 1981; Cant and Morest 1984; Vater et al. 1985). The AVCN is innervated by the ascending branch of the auditory nerve, whereas the DCN and PVCN are innervated by the descending branch (Fig. 2.9). The branching of auditory nerve fibers forms the first division of the central auditory

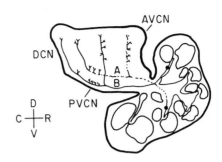

Fig. 2.9. Projections of two auditory nerve fibers to the cochlear nucleus of the horseshoe bat. Notice that each fiber bifurcates into an ascending branch that innervates the anteroventral cochlear nucleus (*AVCN*), and a descending branch that innervates the posteroventral and dorsal cochlear nuclei (*PVCN* and *DCN* respectively). The fibers from the apical part of the cochlea (*A*), represent low frequencies and are distributed in different regions of the cochlear nucleus than are fibers originating from the basal part of the cochlea (*B*), where high frequencies are represented. The dorsoventral and rostrocaudal axes are indicated. (After Vater et al. 1985)

system into prallel pathways, and the manner in which they distribute themselves creates three separate tonotopic organizations, one in each of the three major divisions of the cochlear nucleus.

The tonotopic organization of these multiple areas has been demonstrated in detail in rufous horseshoe bats by tracing the connections of physiologically defined parts of the cochlear nucleus (Feng and Vater 1985). Small deposits of the enzyme horseradish peroxidase (HRP) were placed in one division of the cochlear nucleus. This enzyme is not only transported back to the cells of origin in the spiral ganglion but is also transported throughout all the other branches of the fiber into which it is incorporated. For example, if HRP is injected in a region of the DCN that responds to 78.8 kHz, branches that terminate in the 78.8 kHz regions of AVCN and PVCN are also labeled. By a series of such experiments, Feng and Vater were able to produce a detailed map of the tonotopic arrangements within AVCN, PVCN and DCN. A schematic version of their results is shown in Fig. 2.10. Within each division, place along the cochlea is represented by a slab of cells, with successive slabs stacked along a particular axis of the division. In the AVCN each slab has a more or less dorsoventral orientation, and the slabs are serially stacked along the rostrocaudal axis. Thus low frequencies are represented in the rostral AVCN and higher frequencies in the caudal AVCN. The slabs are not of equal dimensions; those representing the acoustic fovea, about 78–79 kHz in rufous horseshoe bats, are disproportionately large and are further distinguished by having exceptionally sharp tuning curves, as do the filter neurons in the auditory nerve.

The isofrequency slabs in the DCN are oriented along an axis similar to that of the AVCN, running rostrocaudally, but the slabs in the PVCN have a different orientation (Fig. 2.10). In the PVCN the frequency organization is

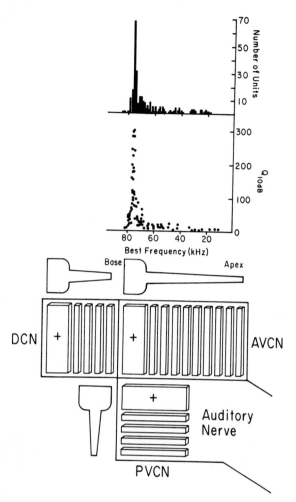

Fig. 2.10. Schematic illustration of the tonotopic organization of the horseshoe bat's anteroventral cochlear nucleus (*AVCN*), posteroventral cochlear nucleus (*PVCN*) and dorsal cochlear nucleus (*DCN*), based on the data of Feng and Vater (1985). Isofrequency contours in each division are represented by *bars,* with the largest bar (+) representing the filter frequencies. A representation of the horseshoe bat's cochlea is above each cochlear nucleus division to illustrate that each division contains a map of the cochlear surface. The distribution of best frequencies and Q_{10dB} values of AVCN neurons is also shown

roughly dorsoventral, with high frequencies dorsal. In this division, as in the AVCN and PVCN, filter units are over represented and have sharp tuning curves. Studies of tonotopy in the mustache bat's cochlear nucleus, although not as detailed as those conducted in the horseshoe bat, have revealed a tonotopic arrangement in each of the three divisions of the cochlear nucleus very similar to that of the horseshoe bat (Kössl and Vater 1985; Ross et al. 1988).

2.4 Superior Olivary Complex

The superior olivary complex is the next group of nuclei in the auditory pathway and is composed of three principal nuclei: the medial superior olive (MSO), the lateral superior olive (LSO) and the medial nucleus of the trapezoid body (MNTB) (Fig. 2.1). Here we only consider the MSO and LSO, since the frequency representation in the MNTB has not been studied in bats. Turning first to the MSO, its tonotopic arrangement was studied by Harnischfeger et al. (1985) in the molossid bat, *Molossus ater*. They found an orderly progression of best frequencies in the MSO, with low frequencies represented dorsally and high frequencies represented ventrally (Fig. 2.11 B). A recent study in mustache bats also demonstrates an orderly tonotopic organization in the MSO as seen from projections to the inferior colliculus (Fig. 2.11 C) (Ross et al. 1986, 1988). The distribution of labeled cells in the

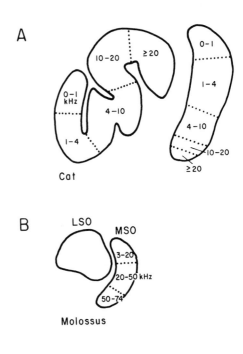

Fig. 2.11 A–C. Tonotopic arrangement of the medial superior olive (*MSO*) and lateral superior olive (*LSO*) in the cat (*A*), in *Molossus ater* (*B*) and in the mustache bat (*C*). The numbers in each region refer to the range of neuronal best frequencies in kHz. Data for the cat are from Guinan et al. (1972 b), for *M. ater*, from Harnischfeger et al. (1985), and for the mustache bat, from Ross et al. (1988)

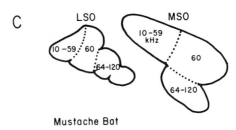

mustache bat's MSO was mapped after HRP was injected into isofrequency regions of the central nucleus of the inferior colliculus, identified with physiological mapping techniques. This study revealed a tonotopic pattern similar to *M. ater,* where low frequencies are located dorsally and high frequencies ventrally. Such a pattern of frequency representation is also seen in the MSO of the cat (Guinan et al. 1972 b) (Fig. 2.11 A) and of the dog (Goldberg and Brown 1968) and is considered to be typically mammalian. The major difference in the mustache bat is that the central portion of the MSO has a thick slab of cells representing 60 kHz, the bat's acoustic fovea.

The tonotopy of the bat's LSO is not as well documented as that of the MSO. Harnischfeger et al. (1985), for example, recorded from only ten LSO cells, a sample size too small to address either the frequency representation or the tonotopic organization. Single unit data on the rufous horseshoe bat show that low frequencies are located laterally and high frequencies medially (Casseday et al. 1988 b). Studies of the connections between isofrequency regions of the inferior colliculus and LSO in the mustache bat are consistent with the single unit data from the horseshoe bat and show that low frequencies are found in the lateral limb of the LSO and high frequencies in the medial limb (Fig. 2.11 C). Sixty kilohertz is over represented in the central region between the high and low frequency representation, as in the MSO. These data suggest a tonotopic pattern, in both horseshoe and mustache bats, that is similar in principle to that of the cat (Tsuchitani and Boudreau 1967; Guinan et al. 1972 b).

2.5 Nuclei of the Lateral Lemniscus

Rostral to the superior olivary complex are the nuclei of the lateral lemniscus. Three major nuclei are recognized in all echolocating bats that have been studied (Figs. 2.1 and 2.12): the ventral nucleus of the lateral lemniscus (VNLL), the intermediate nucleus of the lateral lemniscus (INLL) and the dorsal nucleus of the lateral lemniscus (DNLL). The VNLL and INLL are especially large in echolocating bats. The VNLL can be further divided into two subdivisions, one that has a columnar arrangement of its neurons, the $VNLL_C$, and one that does not (Zook and Casseday 1982 b; Covey and Casseday 1986; Metzner and Radtke-Schuller 1987). The anatomical organization of the $VNLL_C$ and the functional consequences of its tonotopy, are discussed in the next chapter.

Studies of both horseshoe (Metzner and Radtke-Schuller 1987) and mustache bats (Ross et al. 1988) show that each of the nuclei of the lateral lemniscus contains the full range of frequencies audible to the bat, and each has a tonotopic organization in which the filter frequencies are over-represented. The tonotopic organization of the nuclei of the lateral lemniscus in the mustache bat is shown in Fig. 2.12.

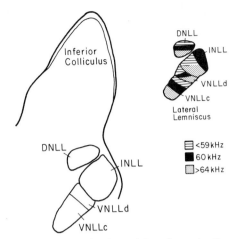

Fig. 2.12. Drawing showing the location of the nuclei of the lateral lemniscus in the mustache bat, and the general tonotopic pattern in each nucleus. Dorsal nucleus of the lateral lemniscus (*DNLL*), intermediate nucleus of the lateral lemniscus (*INLL*), columnar division of ventral nucleus of the lateral lemniscus (*VNLLc*). (Ross et al. 1988)

2.6 Central Nucleus of the Inferior Colliculus

The tonotopy of the inferior colliculus of FM bats conforms to the pattern seen in commonly used laboratory mammals. The general arrangement is that each frequency is represented as a sheet of neurons that receives the projections from corresponding frequency regions of the lower auditory brainstem nuclei (Aitkin 1976). We illustrate this tonotopic arrangement with results from a study of the big brown bat (Covey and Casseday 1986) (Fig. 2.13). The isofrequency sheets in this species are more or less curved slabs of cells that are stacked in an orderly manner along the dorsoventral plane, with low frequencies represented dorsolaterally and high frequencies ventromedially. A similar frequency arrangement has also been found in the inferior colliculus of Mexican free-tailed bats (Bodenhamer and Pollak 1981).

In marked contrast to this common arrangement are the special adaptations in the mustache bat's inferior colliculus (Fig. 2.14), as described by John Zook and his colleagues (Zook et al. 1985). They divided the central nucleus into three main parts of roughly equal size, the anterolateral, medial and dorsoposterior divisions. Physiological recordings in the anterolateral division show that frequency representation changes systematically along the rostrocaudal dimension, from about 10 kHz rostrally to 60 kHz caudally. The isofrequency sheets in the medial division provide an orderly representation of high frequencies, from about 65–120 kHz. The third division, the dorsoposterior division, is unique in that all of its neurons are sharply tuned to a small frequency band around 60 kHz. The dorsoposterior division is the

37

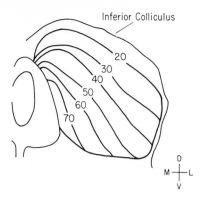

Fig. 2.13. Drawing of a transverse section through the inferior colliculus of the big brown bat showing its tonotopic arrangement. The orientation of isofrequency contours is shown by the *solid lines,* and the frequency in kHz represented in each contour is indicated. (Data from Casseday and Covey)

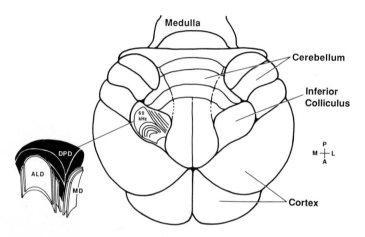

Fig. 2.14. Schematic drawing of a dorsal view of the mustache bat's brain to show the location of the 60-kHz region in the inferior colliculus. The hypertrophied inferior colliculi protrude between the cerebral cortex and cerebellum. In the left colliculus are shown the isofrequency contours, as determined in anatomical and physiological studies. A three-dimensional representation of the laminar arrangement in the inferior colliculus is shown on *far left.* Low frequency contours, representing an orderly progression of frequencies from about 59 kHz to about 10 kHz, fill the anterolateral division (*ALD*). High frequencies, from about 64 kHz to over 120 kHz, occupy the medial division (*MD*). The 60-kHz isofrequency contour is the dorsoposterior division (*DPD*), and is the sole representation of the filter neurons in the bat's colliculus. (Data from Zook et al. 1985. Drawing from Pollak et al. 1986)

isofrequency region where filter units are represented. It thus represents the acoustic fovea and occupies about a third of the volume of the inferior colliculus.

The tonotopy of the mustache bat's inferior colliculus appears to depart markedly from the more general mammalian plan. However, Zook et al. (1985) proposed that the tonotopic arrangement evolved from a generalized tonotopic plan, where the dorsoposterior division represents a greatly hyper-

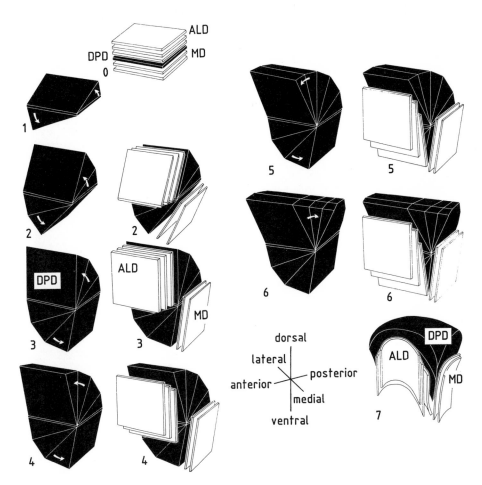

Fig. 2.15. Hypothetical evolution of the dorsoposterior division (*DPD*) from a standard mammalian plan. In the *top figure* is the putative arrangement of isofrequency contours in the mustache bat's ancestor. Each *slab* represents an isofrequency contour, where low frequencies lay dorsally and high frequencies ventrally. Low frequency contours fill anterolateral division (*ALD*), and high frequencies occupy the medial division (*MD*). The "60-kHz" isofrequency contour is the dorsoposterior division (shown in *black*). In the *left*, from *1* to *6* are the progressive axes of expansion: one along the dorsoanterior direction (*arrow on right in 1*) and one along the medioventral axis (*arrow at left in 1*). *Wedges* added in each successive figure indicate expansion of 60-kHz contour during evolution. In the figures on the *right*, the low and high frequency contours are added to complete the picture. The *bottom panel* (*7*) shows the present arrangement of the mustache bat's inferior colliculus. (Zook et al. 1985)

trophied isofrequency contour. They propose that the tonotopic distortion evolved as these bats adopted the use of a Doppler sonar system, which requires a large neuronal population for processing of frequency and amplitude modulations in the 60 kHz echo CF component. A scenario of how this particular tonotopy emerged is illustrated in Fig. 2.15.

Chapter 3

Anatomy of the Auditory Brainstem

3.1 Introduction

Although specialized anatomical features are present in the auditory systems of many echolocating bats, the basic structure of their central auditory system is the same as that of other mammals. In all species a well-defined tonotopic organization is preserved throughout successive levels. The bat differs from other mammals mainly in that its range of hearing extends to unusually high frequencies and that certain frequencies that are important for echolocation are overrepresented. The cochlear nucleus conforms to the general mammalian plan and consists of three major divisions that give rise to divergent projections to the superior olivary complex, to the nuclei of the lateral lemniscus and directly to the inferior colliculus. However, in echolocating bats, certain cell groups within these pathways are unusually large and well differentiated. In the introduction we pointed out that some pathways are designed to provide convergence of signals from the two ears whereas others provide separation of the inputs from the two ears; we refer to these two types of pathways as binaural and monaural, respectively. Virtually all of the brainstem auditory pathways, whether binaural or monaural, converge at a common destination in the inferior colliculus. Thus, in every species, a major challenge in the study of auditory pathways is the question of how signals, relayed via many different parallel routes, interact in the midbrain to produce an output that synthesizes the information from all these sources.

In this chapter, we shall consider how general mammalian auditory structures could function in the specialized process of echolocation; we shall also consider the anatomical specializations that have evolved in bats and their possible significance for echolocation. We shall emphasize the connections of the anteroventral cochlear nucleus; it is not only the largest division of the cochlear nucleus in bats, but its connections provide the clearest clues about function in all mammals. We shall mainly refer to the literature on central auditory pathways in bats; reference to the more general literature on other mammals can be found in these sources and in recent reviews (Warr 1982; Cant and Morest 1984; Casseday and Covey 1987).

3.2 Three Divisions of the Cochlear Nucleus and the Origins of Parallel Auditory Pathways

Earlier we pointed out that the auditory nerve divides into an ascending branch that innervates the anteroventral cochlear nucleus (AVCN) and a descending branch that innervates the posteroventral (PVCN) and dorsal (DCN) cochlear nucleus. Figure 3.1 illustrates the division of the auditory nerve into parallel systems within the cochlear nucleus. Also shown are some features of the cell types and innervation patterns within each of the cochlear nucleus divisions. These features are important for further consideration of parallel pathways.

The ascending branch of the auditory nerve consists of fibers that make successive contacts on different types of cells in AVCN. The entire set of cells that receive input from a fiber forms a sheet or slab that in turn defines the

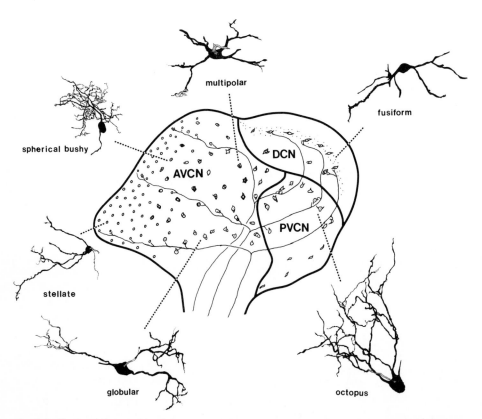

Fig. 3.1. Cytoarchitectural divisions and cell types in the cochlear nucleus. The cochlear nucleus is shown in a parasagittal view with anterior at *left*. Within the nucleus are drawings of Nissl stained cells to show the size and shape of some of the cell types that characterize different parts of AVCN, DCN and PVCN. The *outset drawings* show how the same types of cells appear in Golgi-impregnated material

orientation of an isofrequency contour (Feng and Vater 1985). The distribution of different cell types is more or less orthogonal to the tonotopic contours, so that each contour contains all or most of the different cell types. The different types of cells are distributed in such a way that certain major subdivisions can be recognized in all mammals. The methods that have been used to define cell types in the cochlear nucleus are Nissl stains to show cell bodies (Osen 1969) and Golgi impregnations or filling with horseradish peroxidase (HRP), to show the morphology of the dendrites or axonal endings of a neuron (Brawer et al. 1974; Cant and Morest 1979; Lorente de Nó 1981).

In Nissl stains of the AVCN of the cat, the distinguishing cell type in anterior AVCN is a round cell that Osen (1969) called the spherical cell. Later studies by Brawer et al. (1974) and Cant and Morest (1979) showed that in Golgi material the spherical cells correspond to a particular Golgi type, the bushy cell, so named because it has one large dendrite from which arises a highly branched or "bushy" dendritic tree. In the most anterior part of AVCN the spherical bushy cells, as they are now called, are mixed with cells whose shape in Nissl stains are ovoid (Osen 1969); in Golgi material these ovoid cells have several dendrites extending in a star-like manner, hence the name stellate (Brawer et al. 1974; Cant and Morest 1979).

Figure 3.1 shows that in bats these types of cells are also seen, and their distribution in the anterior AVCN is the same as in other mammals. The most anterior part of AVCN contains a homogeneous population of spherical cells that are like those in the cat except that they are extremely small (Zook and Casseday 1982a). Just posterior to this area ovoid cells are mixed with the spherical cells. In Golgi material we see bushy and stellate cells in this area (Fig. 3.1). In bats, we do not yet know whether the different Nissl types, spherical and ovoid, correspond to the Golgi types, stellate and bushy although it seems likely that they do.

The posterior part of AVCN of the bat also resembles posterior AVCN of the cat in that 1) it has a heterogeneous population of cells, and 2) its cells are of the same types as those in the cat's AVCN. For the present purpose, the most significant cell type is the globular cell. In Nissl stains these cells have an eccentric nucleus (Osen 1969), and in Golgi material they have bush-like dendritic fields (Brawer et al. 1974). In bats, the dendritic bush may not be as prominent as in the cat, but the other features, eccentric nucleus and projections, are the same. Later we shall describe how the spherical bushy cells and the globular cells are related to binaural processing. Another type of cell that is common in posterior AVCN is the multipolar cell. In the bat these are the largest cells in AVCN, and they are characterized by several thick dendrites that radiate in different directions.

The idea that these different cell types reflect functional differences is reinforced by the fact that they receive different types of endings. The bushy spherical cells and the globular cells are especially well suited to receive and transmit temporal information because they receive very large endings from auditory nerve fibers, the end bulbs of Held (Fig. 3.2). The presence of such a large and secure synaptic arrangement suggests a one-to-one transmission of

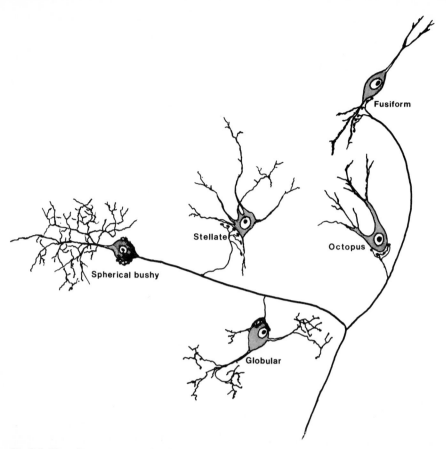

Fig. 3.2. Drawing to summarize how an auditory nerve bifurcates and ends on different types of cells in the cochlear nucleus. The ascending branch is shown ending on a globular cell, stellate cell, and spherical bushy cell. Note the end bulbs of Held on the spherical bushy cell and the globular cell. The descending branch is shown ending on the cell body of an octopus cell and on the basal dendrites and cell body of a fusiform cell. The view is parasagittal, as in Fig. 3.1, with anterior at left and dorsal at top

information from the ending to the post-synaptic cell, so that in this part of the pathway, temporal information is preserved (Pfeiffer 1966; Morest 1968; Bourk 1976; Rhode et al. 1983; Rhode 1985; Smith and Rhode 1987). In the cat, the low frequency fibers provide more (Ryugo and Rouiller 1988) and larger end-bulbs to AVCN bushy cells (Rouiller et al. 1986) than do high frequency fibers; it is at the lower frequencies that temporal information must be preserved for the binaural processing that occurs at the superior olive. Later we shall point out that the cell types are, in part, the basis for subdivision into different pathways to the superior olives. First we briefly describe the two other major divisions of the cochlear nucleus.

Because the functions of PVCN and DCN are not well understood in the bat or any other mammal, we shall only make brief comments on the struc-

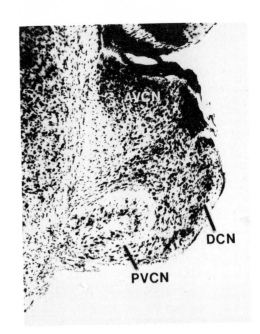

DCN

PVCN

Fig. 3.3. Photomicrograph of a horizontal section through the posterior part of the cochlear nucleus in a horseshoe bat. In the PVCN of this bat the octopus cell area forms a distinct horseshoe shape. The functional significance of this configuration is unknown

ture of these nuclei. First, the structure of PVCN is clearly different from that of DCN, and both differ from AVCN. The PVCN contains small multipolar and small elongate cells, but the most distinct group of cells are the so called octopus cells (Osen 1969). These cells have two or more large dendrites that extend perpendicularly to the fibers of the descending root of the auditory nerve (Figs. 3.1 and 3.2). In most bats the octopus cell area is prominent, and in some horseshoe bats the octopus cells are embedded within the fibers of the descending root in a remarkably elegant formation (Fig. 3.3). We do not yet know the functional significance of this structural specialization.

In many mammals the chief characteristic of DCN is the laminar arrangement of its cells. At the surface is a superficial layer of fibers and granular cells; below this is an intermediate layer of fusiform and small cells, and below this, a deep layer containing a mixture of cell types, including giant multipolar cells. The shape and position of the fusiform cells distinguishes them from other cells in the cochlear nucleus. They are aligned in a single layer with their long axes perpendicular to the surface of the nucleus, so that one dendrite extends to the deep layer and the other to the superficial layer. The fusiform cells are especially interesting in that their basal dendrites receive input from the auditory nerve, whereas their apical dendrites receive input from AVCN (Jones and Casseday 1979). These connections suggest that DCN integrates some functions of the auditory nerve with functions of AVCN; however, we have little information about the nature of this integration. In bats, the DCN is the smallest division of the cochlear nucleus. In some bats such as the mustache bat, the lamination is poorly developed. In the horseshoe bat, the part that represents frequencies below the filter frequencies is laminated; the part that represents frequencies at and above the filter frequencies is not (Feng and Vater 1985). We do not yet know the significance of this arrangement, but it is clearly a specialization that is species-specific rather than specific to the long CF bats, because the mustache bat does not have it.

Each of the three main divisions of the cochlear nucleus provides one or more separate parallel pathways to the auditory centers in the medulla and midbrain. Some of these pathways are monaural, others binaural. The pathway from the DCN is an example of one that is exclusively monaural (Fig. 3.8). Axons of fusiform cells and cells in the deep layers of DCN project to the contralateral inferior colliculus, bypassing the binaural centers of the superior olivary complex (Schweizer 1981; Zook and Casseday 1982b, 1985). From PVCN, axons of octopus cells and other cells also bypass the binaural centers and project contralaterally to innervate the ventral and intermediate nuclei of the lateral lemniscus and the inferior colliculus (Zook and Casseday 1982a, 1985). The major olivary targets of PVCN are the periolivary nuclei, the small aggregates of cells clustered around the binaural nuclei. Many of these periolivary cells project back to the cochlea or to the cochlear nuclei, making the PVCN an important link in the efferent auditory pathways.

In contrast to DCN and PVCN, the functions of AVCN are relatively well understood. Therefore, we devote the next section to a description of the connections of AVCN and the probable significance of these connections for function.

3.3 Binaural and Monaural Pathways of the AVCN System

The AVCN is the source of three major systems of pathways; one of these is binaural and the other two are monaural: 1) binaural pathways to MSO, LSO and MNTB, the main nuclei of the superior olivary complex that together form the major pathway in the brainstem for convergence from the two ears (Figs. 3.4 and 3.6); 2) direct projections to the monaural divisions of the contralateral nuclei of the lateral lemniscus, the INLL and the VNLL (Fig. 3.8); 3) direct projections to the contralateral inferior colliculus (Fig. 3.8). The lower brainstem targets of AVCN also project to the inferior colliculus and thus may be considered as part of a multisynaptic pathway from AVCN to the inferior colliculus. We refer to the entire set of pathways that originate in AVCN and project directly or indirectly to the inferior colliculus as the "AVCN system".

3.3.1 Binaural Pathways and the Localization of Echoes

A very important part of echolocation is the localization of sound source. It has been recognized since the early work of Stotler (1953) that the binaural

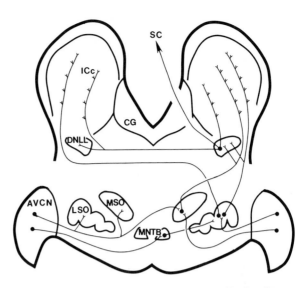

Fig. 3.4. Summary diagram to show ascending binaural pathways. To simplify the picture only projections via the right superior olivary complex are shown. The MSO receives direct projections from the AVCN's of both sides. The LSO receives direct projections from the ipsilateral AVCN and indirect projections, via MNTB, from the contralateral AVCN. The MSO projects ipsilaterally to DNLL and to the inferior colliculus. The LSO projects bilaterally to DNLL and to the inferior colliculus. The DNLL projects bilaterally to the inferior colliculus and also projects to the deep layers of the superior colliculus (*SC*). (Casseday and Pollak 1989). Abbreviation: *CG* central gray

connections of the superior olivary complex must play in important role in sound localization (Casseday and Covey 1987). Because the beginning of the pathway is the AVCN, we start by examining the types of cells that project to MSO, LSO and MNTB. The main components of the binaural pathways are shown in Fig. 3.4, and photomicrographs of the superior olive of the mustache bat are shown in Fig. 3.5.

The binaural inputs to the MSO come almost exclusively from spherical bushy cells in the ipsilateral and contralateral cochlear nuclei (Fig. 3.6 A). The binaural projections to LSO are slightly more complex (Fig. 3.6 B). It is also innervated by spherical cells, probably the same population of cells that innervate the MSO. However, LSO is directly innervated only by the ipsilateral AVCN. The contralateral innervation arises via an intermediate synapse in the MNTB (Harrison and Warr 1962; Zook and DiCaprio 1988 a, b). The source of projections to MNTB is the population of globular cells in posterior AVCN. The axons of the globular cells are some of the largest in the auditory pathway, and their terminations, the calyces of Held, are the largest terminal specializations found in the mammalian brain (Nakajima 1971; Cant and Morest 1984; Zook and DiCaprio 1988 b). Figure 3.7 shows a calyx of the type usually found on principal cells in MNTB. These endings are similar to but larger than the end bulbs of Held that innervate spherical and globular cells. These pathways are apparently specialized for rapid and

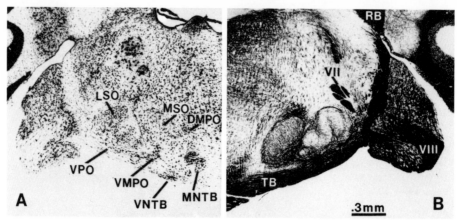

Fig. 3.5. A, B. Photomicrographs of the superior olivary complex of the mustache bat. A Section through the left brain stem stained for cell bodies (Nissl stain). B Section through the right side of the brain stem, at the same level as shown in A, stained to show fibers. The dense fiber plexus reveals the shape of MSO and LSO. Abbreviations: *DMPO* dorsomedial periolivary nucleus; *MNTB* medial nucleus of trapezoid body; *RB* restiform body; *TB* trapezoid body; *VII* 7th nerve; *VIII* 8th nerve; *VMPO* ventromedial periolivary nucleus; *VNTB* ventral nucleus of trapezoid body; *VPO* ventral periolivary nucleus. (Zook and Casseday 1982a)

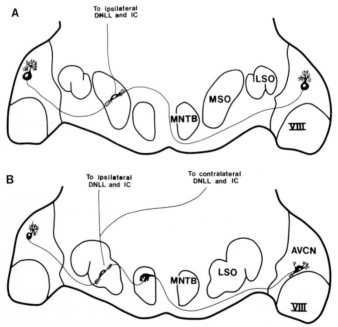

Fig. 3.6 A, B. Diagrams to show projections to MSO (A) and LSO (B). A The bipolar cells of MSO receive input from the ipsilateral ear on the lateral dendrite and input from the contralateral ear on the medial dendrite. In both cases the source of these inputs are spherical bushy cells in anterior AVCN. B The cells in LSO receive input from ipsilateral AVCN, from spherical bushy cells as well as from globular cells in posterior AVCN that project to MNTB and terminate on the cells in MNTB as calyxes. The cells in MNTB then project to LSO. See Fig. 3.5 for abbreviations

48

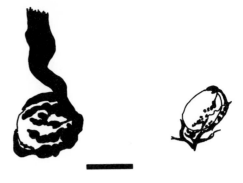

Fig. 3.7. End bulbs on cell in columnar VNLL (*right*) and calyx on cell in MNTB (*left*). The endings in VNLL are slightly smaller than the end bulbs found in the anterior AVCN. The calyxes in MNTB are the largest endings found in the mammalian brain. Drawn from fibers filled with HRP in big brown bat. Calibration bar = 10 μm

secure transmission of information from the two ears to cells in MSO and LSO.

The efferent projections of LSO and MSO define the ascending course of the binaural pathways (Figs. 3.4 and 3.6). These travel in the lateral lemniscus, bypass the intermediate and ventral nuclei of the lateral lemniscus, and innervate the dorsal nucleus of the lateral lemniscus (DNLL) and the central nucleus of the inferior colliculus. Thus in the lateral lemniscus, DNLL is the only nucleus included in the binaural system. As we show below, the projections from MSO, LSO and DNLL have widespread targets in the central nucleus of the inferior colliculus, and thus much of the inferior colliculus is included in the binaural system.

The ascending projections of DNLL are significant because it is here that we first see a connection with centers for motor action. In addition to its projections to the inferior colliculus, the DNLL projects to the deep layers of the superior colliculus (Covey et al. 1987), a motor center for orienting the head and pinnae. Thus the DNLL is clearly one source by which information about sound location can elicit the behavior of orienting toward sound. We shall return to this point when we describe auditory pathways that bypass the inferior colliculus.

3.3.2 Functional Considerations of LSO and MSO Pathways

The difference in the innervation patterns of MSO and LSO has functional significance. The direct input to MSO from both sides provides essentially the same type of information from either ear, and it turns out that many cells here are excitatory from both sides (Goldberg and Brown 1968; Caird and Klinke 1983; Harnischfeger et al. 1985). The indirect input to LSO from the contralateral side allows a transformation of the signal, and thus most LSO neurons are excited by the direct input from the ipsilateral ear but are inhibited by the indirect input from the contralateral ear (Boudreau and Tsuchitani 1968; Guinan et al. 1972a; Harnischfeger et al. 1985; Casseday et al. 1988).

Figures 3.4 and 3.6 illustrate an important difference between the ascending projections of LSO and those of MSO. The MSO projects ipsilaterally to the inferior colliculus, whereas the LSO projects bilaterally. Because we know the projections and the response properties of LSO and MSO neurons, one suspects that we might be able to predict, from a strictly connectional point of view, what classes of binaural cells should be found in the inferior colliculus. For example, from the MSO input some cells should be excited by sounds presented to either of the two ears (E-E cells), and such cells are common. From the LSO, which projects bilaterally, we would expect the number of cells that are inhibited by the ipsilateral ear and excited by the contralateral ear (E-I) to be in equal proportion to those that are excited by the ipsilateral ear and inhibited by the contralateral ear (IE). In fact the latter expectation is not fulfilled; there are very few I-E cells in the inferior colliculus.

This puzzling finding might be explained by recent studies which suggest that the crossed and uncrossed projections from the LSO to the inferior colliculus employ different transmitters (Hutson et al. 1987; Saint Marie et al. 1988). The uncrossed projections are glycinergic, and probably provide inhibitory inputs to the colliculus, whereas the crossed projections most likely provide excitatory input to the colliculus. Thus the crossed projection from I-E cells of the LSO should evoke E-I properties in the inferior colliculus neurons on which they synapse. The influence of the uncrossed projection would be more difficult to discern, since they presumably provide only inhibition to their post-synaptic targets in the inferior colliculus.

3.3.3 Do All Bats Have an MSO?

When one considers the cues that are used in localization, it might seem surprising that an animal as small as a bat has an MSO. Large land-dwelling mammals are sensitive to low frequency signals, use binaural *time* cues for localizing sound and have large MSOs; small mammals use high frequency signals, use binaural *intensity* cues, have a large LSO but often have no MSO (Masterton et al. 1985). Thus it is not surprising that LSO is large and prominent in bats; early reports that MSO is absent in bats are consistent with this idea (Harrison and Irving 1966). Later it was discovered (Zook and Casseday 1982a) that the mustache bat has a rather large structure that looks like MSO (Fig. 3.5). This finding opened the issue of whether bats have an MSO and, if so, what might be its functional significance. The first part of this question has been addressed by connectional studies that show that the MSO in the mustache bat is like that of other mammals: It receives bilateral input from anterior AVCN (Zook and Casseday 1985), and it projects to the ipsilateral inferior colliculus (Zook and Casseday 1982b, 1987). We have no answer to the question of functional significance, but comparative observations may eventually provide a clue. For example, we have examined the superior olivary complex of a large number of echolocating bats. The results suggest that the LSO is a cell group that varies little among most species. However the cell groups medial to LSO vary considerably, in cytoarchitecture and size: Some bats have a "typical" MSO, others have none, and others have two structures, neither of which has all the features of MSO. The conclusion is that these medial cell groups appear to have been more plastic in evolution, but for what purpose remains to be seen (Casseday et al. 1988).

3.3.4 Monaural Pathways

We turn now to those pathways from AVCN that bypass the superior olives to terminate in the contralateral nuclei of the lateral lemniscus and in the inferior colliculus (Fig. 3.8). Little is known of the function of these pathways in mammals. The unusual development of the lateral lemniscus in echolocating bats has led to experiments, described below, that provide some clues about the function of these pathways.

In echolocating bats, the VNLL and INLL are especially large and well organized (Figs. 3.9 and 3.10). Their structures are similar in all bats that have been studied, even those that use markedly different sonar signals, such as the mustache bat and the big brown bat. For example, in both species, half of the VNLL has a columnar organization in which the cells are organized in strings or columns parallel to the ascending fibers of the lateral lemniscus (Figs. 3.9 and 3.10). Furthermore, the connections of INLL and VNLL are esentially the same in these two species. These observations, that there are constant features in the structure of the nuclei of the lateral lemniscus, may indicate that the functions of these nuclei are to process features of the sonar signals common to both bats. The most obvious feature in common is the brief FM component.

The cytoarchitectural divisions in the nuclei of the lateral lemniscus have a counterpart in the projections from AVCN. Figure 3.9 A shows that the projection from AVCN diverges to three separate targets, one in INLL and

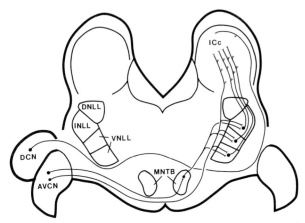

Fig. 3.8. Schematic drawing to show some of the monaural pathways in the brainstem of mustache bat. DCN projects directly to the contralateral inferior colliculus. AVCN projects directly and indirectly via relays in MNTB and in the nuclei of the lateral lemniscus. Pathways from PVCN are not shown because of insufficient knowledge about their course and terminations in the mustache bat, but they do contribute to the monaural pathways. (Casseday and Pollak 1989)

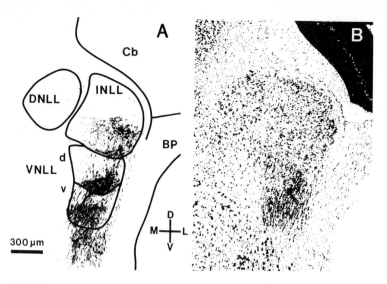

Fig.3.9 A, B. Ascending projections from AVCN to the lateral lemniscus and cytoarchitecture of nuclei of the lateral lemniscus in the mustache bat. *A* Projections from the posterior part of AVCN diverge to three main targets, one in INLL and one in each part of VNLL. The photomicrograph is a negative image of the darkfield view of an autoradiograph to show transport of [³H]-leucine from AVCN. *B* The same section as shown in *A* is stained to show the different divisions of the lateral lemniscus in this photomicrograph. (Zook and Casseday 1985). Abbreviation not in text: *BP* brachium pontis; *Cb* cerebellum; *d,v* dorsal and ventral divisions of VNLL

two in VNLL. The INLL and VNLL also receive inputs from MNTB (Fig. 3.8). Thus each of these nuclei in the lateral lemniscus receives more than one type of monaural input from the contralateral ear: 1) direct input from AVCN and 2) indirect input via a synapse in MNTB. It is noteworthy that many of the afferent endings on cells in the columnar VNLL are calyciform synapses (Figs. 3.7 and 3.12). These endings are similar to the calyciform endings in AVCN and in MNTB, the two nuclei that provide the major afferent input to the columnar VNLL.

Such a system of parallel inputs provides a mechanism by which an auditory signal could undergo different transformations, including different delays of arrival. Each division of the lateral lemniscus in turn sends a projection to the ipsilateral inferior colliculus (Schweizer 1981; Zook and Casseday 1982b; Covey and Casseday 1986; Ross et al. 1988).

The connections from AVCN to the columnar division of VNLL and from there to the inferior colliculus are especially useful in gaining clues about structure-function relationships. The columnar organization of this nucleus can be seen in the dorsoventral dimension (Fig. 3.10). A second organization is revealed in studies of axonally transported tracers to show afferent and efferent connections of the columnar VNLL (Covey and Casseday 1986).

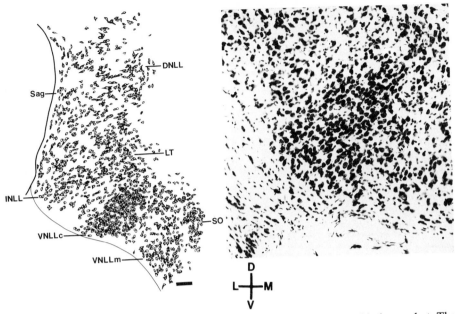

Fig. 3.10. Cytoarchitecture of the nuclei of the lateral lemniscus in the big brown bat. The drawing at the *left* shows all the nuclei of the lateral lemniscus and surrounding cell groups. In the photomicrograph at the *right,* the columnar VNLL is the triangular area of darkly stained cells. The columnar area (*VNLLc*) is easily identified by its small and densely packed cells and by the beaded or columnar arrangement of the cells along the length of the ascending fibers. Abbreviations not used in other figures: *LT* lateral tegmentum; *Sag* sagulum; *SO* superior olive; *VNLLm* multipolar cell area of VNLL. (Modified from Covey and Casseday 1986)

In the experiments to show afferent connections to the columnar VNLL, the tracer wheat germ agglutinin conjugated to horseradish peroxidase (WGA-HRP), was injected in part of AVCN. The WGA-HRP is carried by anterograde axonal transport to its targets. The result in columnar VNLL is a thin band of labeled fibers that extends mediolaterally, at right angles to the columns of cells. If the WGA-HRP is injected in a high frequency area of AVCN, the band of transport is in the ventral part of columnar VNLL; if the injection is in a low frequency area, the band is dorsal. When these projections are traced through the anterior-posterior dimension, we see that they form a thin *sheet* of fibers. The entering fibers innervate a sheet one or two cells in thickness. The location of these sheets of projections in the columnar VNLL was more closely related to the tonotopic locus of the injection than were the projections to other auditory nuclei.

In studies to show efferent connections from the columnar VNLL to the inferior colliculus, the same sheet-like arrangement was seen. In these experiments, the tracer was injected in a known frequency region of the inferior colliculus, and the retrograde transport to the columnar VNLL marked the cells that project to the area of the injection. The results showed that labeled cells were always aligned in a sheet. These sheets of labeled cells had the same rela-

53

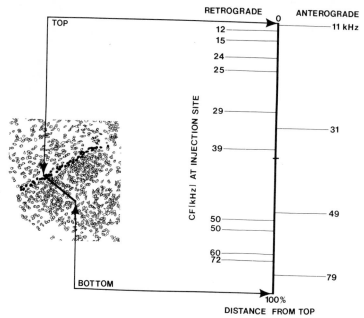

RETROGRADE · ANTEROGRADE

TOP · BOTTOM

CF(kHz) AT INJECTION SITE

DISTANCE FROM TOP

Fig.3.11. Diagram to summarize the relationship between the tonotopy of the injection site and the location of anterograde or retrograde transport in the columnar area. Distance through the nucleus (*left*) is shown as percent from top to bottom (*right*). Along this scale is shown the relative location of the sheets of retrograde or anterograde transport. The *numerals* indicate the best frequency (*CF*) in kHz of cells at the injection site. The *position* of each numeral along the vertical axis indicates the position of the sheet of label as a proportion of the distance from the dorsal border of the columnar area. At the *left* of the line is shown retrograde transport from the inferior colliculus and at the *right*, anterograde transport from the AVCN. For example, cells shown in *black* depict the position of a sheet of cells labeled by an injection of HRP in a low frequency (11 kHz) region of the inferior colliculus. Lateral is to left. (Covey and Casseday 1986)

tion to the tonotopic organization of the inferior colliculus as did the anterograde sheets to the tonotopic organiziation of the AVCN. This relationship is summarized in Fig. 3.11, and an example of a fiber innervating one row is shown in Fig. 3.12. Sheets of cells in the dorsal part of the columnar VNLL receive projections from low frequency parts of AVCN and project to low frequency parts of the inferior colliculus; sheets of cells in the ventral part receive projections from high frequency parts of AVCN and project to high frequency parts of the inferior colliculus. The entire range of frequencies audible to the bat is systematically represented along the dorsoventral dimension of the columnar VNLL. However, there is an overrepresentation of frequencies between 25–50 kHz, corresponding roughly with the range of the FM echolocation call emitted by the big brown bat. It is important to see that each column is only approximately 20–30 cells in height (Figs. 3.10 and 3.11). This observation must mean that frequency representation is compressed in this dimension.

54

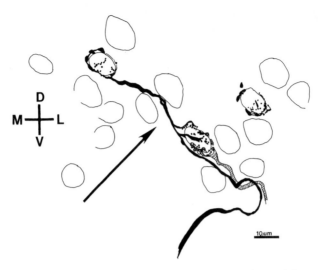

Fig.3.12. Camera lucida drawing of fibers and endings in the columnar area after an injection of HRP in AVCN. Two fibers turn medially from the lateral lemniscus. One fiber (*black*) terminates on two cells in a row of the VNLLc. A second fiber (*striped*) terminates on one of these cells. Endings of a cell in another row are shown, but the fiber could not be traced. The *arrow* indicates the orientation of the columns. (Covey and Casseday 1986)

The exact nature of the frequency compression in the columnar VNLL is not known. The authors suggested two possibilities that have functional implications for processing an FM sweep: 1) only selected frequencies are processed in the columnar VNLL, or 2) each sheet receives convergent projections from a band of frequencies. The result in either case is a sort of comb filter. The effect of such a "filter" in processing an FM signal would be to cause cells in one sheet to respond almost simultaneously to a particular frequency or to a narrow band of frequencies; the cells in an adjacent sheet would respond over a different frequency or band of frequencies. These two frequency bands would be separated by a temporal gap due first to their order of occurrence in the FM signal and, second, to differences in travel time for different frequencies along the basilar membrane. Thus, in response to an FM stimulus, the combined output of all the sheets in the columnar area would convert the smooth FM sweep into a series of temporally discrete responses that mark the time of occurrence of specific frequency bands. Related to this hypothesis of temporal processing is the observation that the afferent endings on cells in the columnar VNLL are calyciform synapses (Figs. 3.7 and 3.12). As mentioned previously, these features suggest a precise, faithful transmission of temporal information. In Chapter 4 we will relate this system of connections to temporal processing in the inferior colliculus and particularly to units that respond with constant latency regardless of the intensity of the stimulus.

3.4 Convergence and Integration at the Inferior Colliculus

The problems in analyzing the inferior colliculus must be viewed in the light of principles given earlier: 1) Each nucleus in the auditory pathway is tonotopically organized and projects to successively higher nuclei in a tonotopic fashion. 2) There are a number of parallel auditory pathways in the lower brainstem. 3) Most, if not all, of these separate pathways represent functional subdivisions in that each is specialized to extract a different type of information from an auditory signal; we have suggested two basic functional divisions, monaural and binaural systems.

The inferior colliculus presents an especially interesting challenge for understanding how the ascending auditory system processes sound. If the system divides the processing into separate tasks in the lower brainstem, then to a large extent it merges the outcomes of these tasks again at the inferior colliculus. That is, many of the pathways that arise separately, from the cochlear nuclei, superior olivary nuclei and nuclei of the lateral lemniscus, converge at the inferior colliculus. Our purpose in this section is to present evidence from studies of the mustache bat to show that although there is convergence of many pathways at the inferior colliculus, some separation of these pathways also occurs in parts of the central nucleus. The evidence for this separation and convergence of pathways is best seen in the 60 kHz isofrequency region of the mustache bat's inferior colliculus. The patterns of convergence and divergence in this region, together with the evidence on the laminar distribution of ascending input to other isofrequency contours of the central nucleus, provide insight into the organization of ascending pathways to the inferior colliculus.

3.4.1 Intrinsic Organization of the Inferior Colliculus

Figure 3.13 shows photomicrographs of Nissl stained sections of the inferior colliculus of the mustache bat. Although there are cytoarchitectural divisions within the inferior colliculus, they provide little insight into the function of neurons within the colliculus or into the way in which each division is related to the tonotopic map. Therefore, a different level of analysis is necessary. In studies of the cat's inferior colliculus, Golgi methods have been useful in showing patterns of dendritic and axonal orientation (Morest and Oliver 1984; Oliver and Morest 1984). As seen with these methods, the principal cell type in the central nucleus is a bipolar, fusiform neuron, the so called disc-shaped cell (Figs. 3.14 and 3.15). Within a given zone of the inferior colliculus the dendrites of nearby cells all have more or less the same orientation, creating a planar organization or laminar pattern. Small and large stellate cells are also present in the central nucleus, but these do not contribute to the planar organization.

Zook et al. (1985) used Golgi methods to analyze the neuronal architecture and orientation of the disc-shaped cells in the inferior colliculus of the

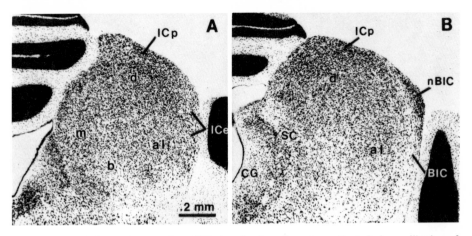

Fig. 3.13 A, B. Photomicrographs of Nissl stained sections from the inferior colliculus of mustache bat. The central nucleus can be divided into dorsal (*d*), anterolateral (*al*) and medial (*m*) parts on the basis of the relative distribution of different types of cells, although the borders between the divisions are not sharp. The clearest border is a cell sparse band (*b*) containing large cells between (*m*) and (*al*). (Zook and Casseday 1982a). Abbreviations not found in text: *BIC* brachium of inferior colliculus; *CG* central gray; *ICc,p* Inferior colliculus, central or pericentral divisions; *nBIC* nucleus of brachium of inferior colliculus; *SC* superior colliculus

mustache bat. According to the dendritic orientation, they divided the central nucleus into three main parts of roughly equal size: the anterolateral, medial and dorsoposterior divisions (Fig. 3.14). These divisions correspond closely to those previously described from Nissl stained material (Fig. 3.13) (Zook and Casseday 1982a). Figures 3.14 and 3.15 show that in two of these divisions, the anterolateral and the medial, the disc-shaped cells are most striking. The result is a laminar-like appearance much like that of the inferior colliculus of other mammals. The third part, the dorsoposterior division, is also characterized by disc-shaped neurons. The disc-shaped neurons in the dorsoposterior division are similar to, but slightly smaller than, those in the other divisions and with less of a tendency for the dendrites to have a common orientation. The dorsoposterior division also has a higher concentration of small stellate (star-shaped) cells than the other divisions. The result is an apparent lack of laminar organization, and this observation was a clue that the dorsoposterior division consisted of one large lamina rather than several small ones.

We pursue this idea by pointing out that the planes or laminae are apparently the anatomical correlates of the isofrequency contours shown in Fig. 2.14 of the previous chapter. Zook et al. (1985) mapped the best frequencies of neurons in the inferior colliculus. In the anterolateral division they found an orderly progression of best frequencies from 10 to 60 kHz along the same rostrocaudal dimension that the laminae are oriented. Likewise, in the medial division, an orderly progression of high best frequencies, from about 64–120 kHz, corresponds to the lateromedial orientation of the laminae.

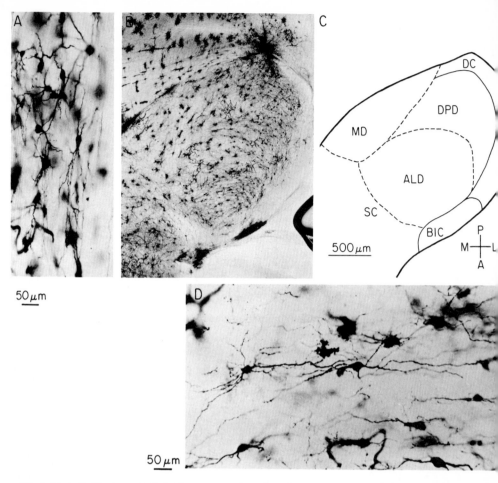

Fig. 3.14 A–D. Horizontal Golgi-impregnated section showing the neuronal architecture of the mustache bat's inferior colliculus. *A* Photomicrograph of disc-shaped neurons in the medial division. *B* Low power photomicrograph showing the prominent laminar arrangement of the fibrodendritic plexuses in the ALD and MD. Note that ALD and MD plexuses are oriented in different planes. DPD laminae are hardly apparent. *C* Drawing to show how the subdivisions of the inferior colliculus correspond to the photomicrograph in *B. D* Photomicrograph of disc-shaped cells in the anterolateral division. Note the interdigitation of the dendrites of adjacent cells. Abbreviations: *ALD* anterolateral division; *BIC* brachium of the inferior colliculus; *DC* dorsal cortex of the inferior colliculus; *DPD* dorsoposterior division; *MD* medial division; *SC* superior colliculus

However, in the dorsoposterior division, only the filter frequencies are represented, i.e., at and around 60 kHz. Thus this division seems to be a greatly hypertrophied isofrequency contour. We shall describe the connections of the dorsoposterior division after we describe, in general, how the separate pathways in the lower brainstem project to the entire central nucleus of the inferior colliculus.

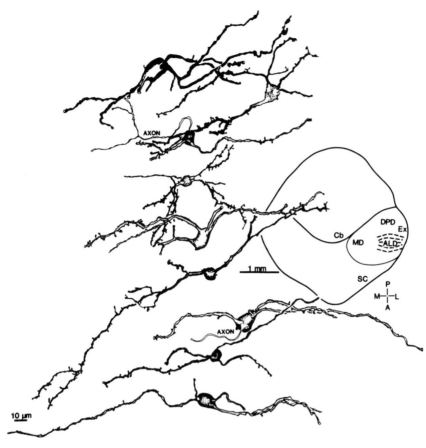

Fig. 3.15. Neuronal architecture of the ALD in the horizontal plane, illustrating the lateral-to-medial gradient in laminar architecture. The dendrites of these cells are oriented more or less mediolaterally. The orientation of the dendrites is indicated in the drawing of the dorsal surface of the brain at the right. Abbreviations: *Cb* cerebellum; *Ex* external nucleus of the inferior colliculus. Other abbreviations are as in Fig. 3.14. (Zook et al. 1985)

3.4.2 Convergence and Divergence of Pathways at the Inferior Colliculus

Figure 3.16 summarizes the results of several experiments (Zook and Casseday 1982a; 1985; 1987) to show connections from the lower brainstem centers to the three divisions of the mustache bat's inferior colliculus. We also show the relation of the projections to the aural areas of the dorsoposterior division that are described below. The first point is that there is a difference between the AVCN target and the DCN target. The projections from AVCN are most dense in the ventral two-thirds of the nucleus, and they do not extend into some of the dorsal parts of the central nucleus. In contrast, the DCN projects in a more widespread manner, more diffusely than

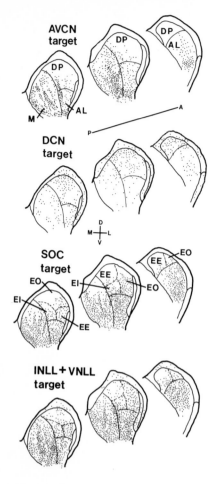

Fig. 3.16. Drawings of anterograde transport of [³H]-leucine in the inferior colliculus to compare the targets of AVCN, DCN, the superior olive and the nuclei of the lateral lemniscus. At the *top* is a summary of transport to the central nucleus of the inferior colliculus following injections in AVCN. Note the absence of transport to most of the dorsal one-third of the central nucleus. The second set of sections shows transport to inferior colliculus from an injection in DCN. Note that there is evidence for transport from DCN to the dorsal parts of the inferior colliculus, that is to those areas that are not labeled after injections in AVCN. The third set of sections summarizes the picture of transport to the central nucleus from the superior olive (SOC). The fourth row of sections shows transport from the ventral and intermediate nuclei of the lateral lemniscus (VNLL and INLL). In the *top row* the three major divisions of the inferior colliculus are labeled: medial division (*M*), anterolateral division (*AL*) and dorsoposterior division (*DP*). In the *third row* we indicate by *dashed lines* our estimate of the location of the aural areas of dorsoposterior division that were defined by Wenstrup et al. (1985). *E-E* refers to the area where neurons receive excitatory inputs from both ears; *E-I* refers to the area where neurons receive excitatory input from the contralateral ear and inhibitory input from the ipsilateral ear; and *E-O* refers to the area where neurons receive input from only one ear, and thus are monaural

AVCN, and it projects most densely to the dorsal areas that are outside the AVCN target.

The second point of Fig. 3.16 concerns the overlap of targets of AVCN with the targets of the superior olive and the nuclei of the lateral lemniscus, which are themselves the targets of AVCN. The projections of the superior olive are coextensive with those of AVCN; again the projections are most dense ventrally and do not extend into the dorsal extremes. The projections of INLL and VNLL are much like those of AVCN and SOC. For the moment we shall ignore the commissural connections of one inferior colliculus to the other, but we shall see later that the inter-collicular connections are important for producing at least one type of binaural response.

These connections have functional implications. Because the most dorsal part of the central nucleus receives input mainly from the contralateral ear, we would expect it to be predominantly monaural in its response properties. In the ventral two-thirds of the central nucleus we would expect a mixture of response properties to reflect a mixture of monaural and binaural inputs: monaural from AVCN, VNLL and INLL, binaural from LSO, MSO and DNLL. In short, this part of the inferior colliculus is mainly the target of pathways that arise directly from AVCN or indirectly from AVCN via synapses in the superior olive or lateral lemniscus. A small monaural contribution also arises from DCN.

The picture now is one of extensive overlap of multiple pathways, especially in the ventral two-thirds of the central nucleus. If we examine the spatial topography of inputs to individual laminae from known sources, e.g., superior olive or lateral lemniscus, will this picture remain or will evidence for segregation of the pathways emerge? We explore this question further by examining the connections of the dorsoposterior division.

3.4.3 Pathways to the Dorsoposterior Division of Inferior Colliculus in the Mustache Bat

The enlarged 60 kHz region of the mustache bat's inferior colliculus is an excellent model for the analysis of convergence and separation of inputs within one isofrequency contour of the inferior colliculus. Injections of HRP confined to this isofrequency contour (Fig. 3.17) reveal that the dorsoposterior division receives projections from the same set of lower auditory nuclei that project to the entire central nucleus of the inferior colliculus. The projections, however, arise from discrete segments of the various projecting nuclei, each of which presumably represents 60 kHz (Ross et al. 1988). It was from these experiments that the tonotopic organization of the mustache bat's auditory brainstem, presented in the previous chapter, was determined.

We carry the evaluation of the projections to the 60 kHz contour a step further by determining which subset of lower auditory nuclei project to specific regions of the dorsoposterior division. The rationale for these studies is that neurons with particular binaural and monaural response properties are topographically arranged within the dorsoposterior division (Wenstrup

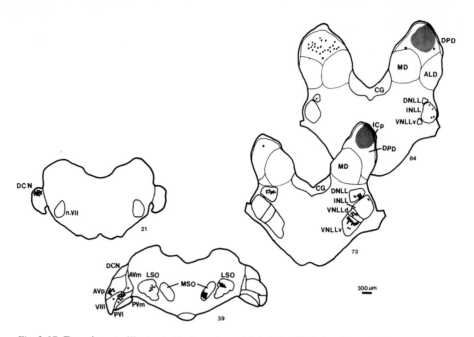

Fig. 3.17. Drawings to illustrate the location of labeled cells following an injection of HRP that was confined to the dorsoposterior division of the mustache bat's inferior colliculus. Before injecting the HRP, the boundaries of the dorsoposterior division were determined by recording the best frequencies of unit clusters with a glass microelectrode. The *hatched region* in the dorsoposterior division indicates the extent of the HRP deposit. All neurons labeled by the retrograde transport of the HRP are shown as *black dots*. Notice that all of the lower nuclei that project to the entire inferior colliculus have labeled cells. Thus, the dorsoposterior division receives the same types of inputs as the entire inferior colliculus. However, the regions that contain labeled cells in each of the lower auditory nuclei are restricted to parts that presumably represent 60 kHz. Abbreviations: *VIII* auditory nerve; *ALD* anterolateral division of inferior colliculus; *AVp* posterior region of anteroventral cochlear nucleus; *AVm* medial region of anteroventral cochlear nucleus; *CG* central gray; *DCN* dorsal cochlear nucleus; *DNLL* dorsal nucleus of the lateral lemniscus; *INLL* intermediate nucleus of the lateral lemniscus; *LSO* lateral superior olive; *MD* medial division of inferior colliculus; *MSO* medial superior olive; *n.VII* nucleus of 7th nerve; *PVl* lateral region of posterocentral cochlear nucleus; *PVm* medial region of posteroventral cochlear nucleus; *VNLLv* columnar division of ventral nucleus of the lateral lemniscus; *VNLLd* noncolumnar division of ventral nucleus of the lateral lemniscus. (Data from Ross et al. 1988)

et al. 1986). Figure 3.16, (third row) summarizes these results. The dorsoposterior division is divided into four aural regions. The monaural neurons are located along the dorsal and lateral parts of the DPD. Neurons that receive excitatory inputs from both ears, E-E neurons, are located in two regions; one in the ventrolateral DPD and one in the dorsomedial DPD. The main population of neurons that receive excitation from one ear and inhibition from the other ear, E-I neurons, is located ventromedially. We will not consider the E-I neurons in the far lateral IC since these are most likely in the external nucleus of the inferior colliculus. Presumably then, the aural properties in each region of the 60 kHz isofrequency contour are a con-

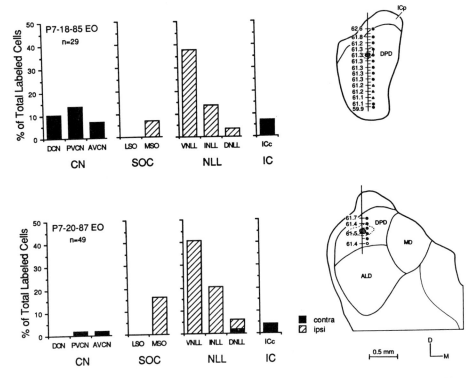

Fig. 3.18. Percentage of labeled cells in different lower auditory nuclei following a small injection of HRP in monaural regions of the dorsoposterior division of the mustache bat's inferior colliculus. The reconstruction of the electrode tracks is shown in the *right*. The best frequencies of the neurons at each location along the electrode track are noted to the *left of each tick mark* along the track. The size of the deposits are indicated by the *small stippled circles,* and were made in central portions of the monaural areas. The percentage of the total number of labeled cells that was found in each nucleus is shown in the *bar graphs on the left.* *Striped bars* refer to labeled cells in ipsilateral nuclei and *black bars* indicate labeled cells in nuclei on contralateral side. In the case shown in the upper panel (P7-18-85 EO), the injection was in the extreme caudal portion of the dorsoposterior division. Note that the largest percentage of labeled cells was found in the VNLL and cochlear nucleus. A small percentage of labeled cells was also found in the MSO, DNLL and opposite inferior colliculus. In the case in the *lower panel* (P7-20-87 EO), the injection was in a much more rostral portion of the dorsoposterior division. The two cases are almost identical in the proportions of labeled cells in nuclei of the superior olive (SOC), lateral lemniscus (NLL) and contralateral inferior colliculus. The cases are different in that many more labeled cells were located in the cochlear nucleus (CN) after the caudal injection than after the rostral injection. Abbreviations for Figs. 3.18–3.21: *ALD* anterolateral division of inferior colliculus; *CG* central gray; *DPD* dorsoposterior division of inferior colliculus; *ICc* central nucleus of inferior colliculus; *ICp* pericentral region of inferior colliculus; *ICx* external nucleus of inferior colliculus; *MD* medial division of inferior colliculus. (Data from Ross and Pollak 1989)

63

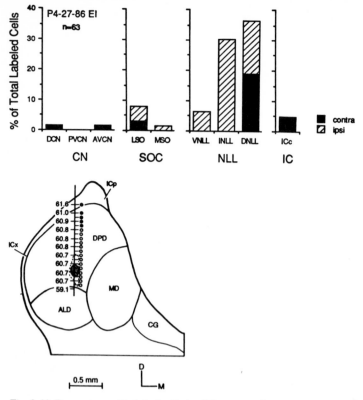

Fig. 3.19. Percentage of labeled cells in different auditory nuclei in the lower brainstem following a small injection of HRP in ventromedial E-I region of the dorsoposterior division of the mustache bat's inferior colliculus. Explanation and abbreviations as in Fig. 3.18. Note the high proportion of labeled cells in the DNLL and INLL. Smaller percentages of labeled cells were seen in the VNLL, opposite inferior colliculus and LSO. See Fig. 3.18 for abbreviations.(Data from Ross and Pollak 1989)

sequence of the projections that terminate in each region. The source of projections to each aural area can be identified with retrograde transport methods by making a small injection of HRP in a region where all neurons were sharply tuned to 60 kHz and have a particular aural response property. With this method, Ross and Pollak (1989) demonstrated that each monaural and binaural region within the dorsoposterior division receives its chief inputs from a different subset of nuclei in the lower brainstem. For most regions, the response properties reflect the subset of inputs.

We consider first the lower nuclei that project to a large area along the dorsal aspects of the dorsoposterior division. This area contains a neuronal population with only monaural response properties. Its location is indicated in Fig. 3.16 (third row) as the E-O area. The top panel of Fig. 3.18 shows the proportion of labeled cells in the various lower nuclei following a small injection of HRP in the posterior and dorsal portion of the monaural area. The largest proportion of labeled cells is in the monaural areas of the lateral lem-

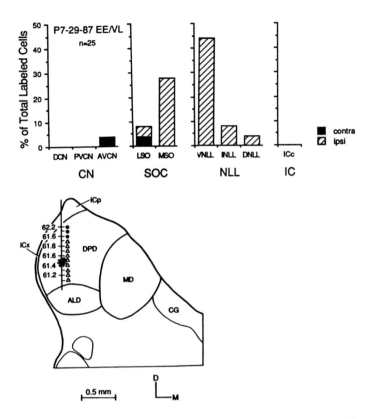

Fig. 3.20. Percentage of labeled cells in different lower auditory nuclei following a small deposit of HRP in lateral E-E region of the dorsoposterior division of the mustache bat's inferior colliculus. Explanation and abbreviations as in Fig. 3.18. Note large percentage of labeled cells in VNLL and MSO. See Fig. 3.18 for abbreviations. (Data from Ross and Pollak 1989)

niscus, particularly VNLL. The cochlear nuclei are other sources of monoaural input. Very little input arises from binaural areas, except for a projection from the MSO, which is surprising. A second case (Fig. 3.18, lowest panel) illustrates that the projections from the MSO were a consistent finding with HRP injections in monaural areas. In this case, the injection was located more anteriorly than in the first case. In both cases there is a small contribution from DNLL, also a binaural nucleus. Even though these binaural inputs are small relative to the inputs from monaural nuclei, it is puzzling that no binaural responses were observed in the region. We do not know why the projections from MSO and DNLL to these areas are reflected in monaural response properties.

The same type of experiment has identified the input to three binaural areas of the dorsoposterior division. Each area has a distinctive pattern of inputs from monaural and binaural nuclei. We first describe the projections to the ventromedial part of the 60 kHz contour, an area in which the neurons

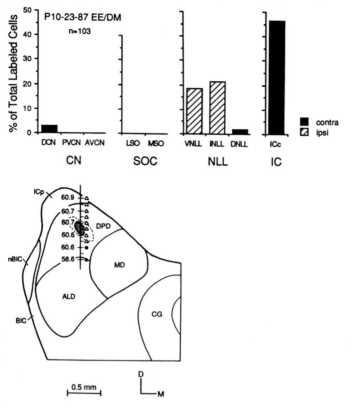

Fig. 3.21. Percentage of labeled cells in different lower auditory nuclei following a small injection of HRP in medial E-E region of the dorsoposterior division of the mustache bat's inferior colliculus. Explanation and abbreviations as in Fig. 3.18. Note the high proportion of labeled cells in the opposite inferior colliculus and the absence of labeled cells in the LSO and MSO. See Fig. 3.18 for abbreviations. (Data from Ross 1989)

are excited by sound to the contralateral ear and inhibited by sound to the ipsilateral ear (E-I cells). The location of this E-I area is shown in the third row of Fig. 3.16. Figure 3.19 shows that the input to this area comes largely from binaural nuclei, especially the DNLL and less so from the LSO. A major input also originates in the INLL. The robust inputs from the DNLL and LSO distinguish the E-I region from the other aural regions of the 60 kHz contour.

The other two binaural areas contain neurons that are excited by sound at either ear (E-E cells). One E-E area is situated laterally and the other medially in the dorsoposterior division as shown in the middle row of Fig. 3.16. Figure 3.20 shows that the two major inputs to the lateral E-E area are from MSO and VNLL. There is little or no input from LSO. This pattern is in marked contrast to projections to the E-I area, which arise from the LSO and from the DNLL and INLL.

Finally, the projections to the medial E-E area (Fig. 3.21) differ substantially from both the E-I region and lateral E-E region. The dominant input

66

arises from the contralateral inferior colliculus. Substantial input also arises from both the VNLL and INLL, but there is almost no input from the binaural centers below the tectum.

These projection patterns suggest that the binaural properties of the medial E-E area arise from interaction via the two colliculi, whereas the binaural properties in the lateral E-E-area arise from the binaural centers in the lower brainstem, especially MSO (Fig. 3.20). In contrast, the binaural properties of the E-I area are shaped largely by the binaural nucleus of the lateral lemniscus, the DNLL (Fig. 3.19).

The connectional differences among the various aural regions of the dorsoposterior division are also reflected in functional differences. Some functional distinctions are apparent, such as between monaural compared to binaural neurons and between E-I and E-E neurons. In the next chapter we discuss in some detail the different manner in which E-E and E-I cells code for sound location. However, our level of understanding does not reach much beyond these more obvious differences. For example, it is clear just from the connectional patterns that there must be at least two major subtypes of E-E neurons, but we have little insight into exactly how the connectional differences are expressed physiologically. Moreover, we presently can only speculate about how the convergence of inputs from several lower nuclei could shape the response characteristics of a collicular neuron expressing a particular aural type. Clarifying these issues represents a major challenge for the future.

3.4.4 Spatial Topography of Inputs to Other Isofrequency Contours

The studies discussed above show a complicated pattern of convergent and divergent projections to regions of an isofrequency contour. Thus, each aural area of the dorsoposterior division receives convergent projections from a subset of lower nuclei. Is there evidence for a similar convergence and divergence of the projections from lower auditory nuclei in other isofrequency contours? A distinctive characteristic of the projections from most lower nuclei is that the ascending axons form prominent bands that are parallel to the isofrequency contours. Figure 3.22 shows the arborization of a single fiber in the inferior colliculus of a big brown bat. The fiber terminates through much of the dorsoventral extent of the nucleus, just as the projection bands (shown in Fig. 3.23), and although it is not shown in the figure, the anterior posterior arborization is almost as extensive. However, the width of the arborization is only about 100 μm. Thus this particular fiber forms a thin sheet that extends throughout much of the inferior colliculus. If most fibers are like this one, we can surmise that each projection band consists of a number of fibers that each arborize only within the thickness of the sheet. It is likely that each sheet is related to a specific isofrequency contour.

Because there is a segregation of pathways within the dorsoposterior division, it is important to determine whether the inputs to other isofrequency contours are also segregated. An important part of the answer would be

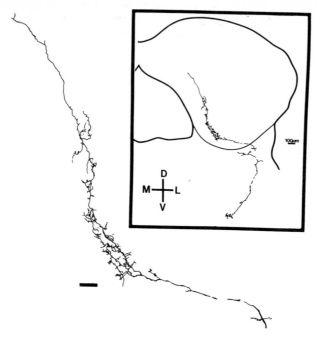

Fig. 3.22. Single fiber filled with HRP ascending in the lateral lemniscus of the big brown bat. The *insert* shows the general orientation of the fiber and its branches with the inferior colliculus. It is most likely that the cell body of origin is the one shown in the columnar VNLL. However, we cannot be absolutely certain because there is a discontinuity in staining, probably at the site of the injection. The width of the region of the greatest branching of the axon is approximately 100 μm, about the same width as the narrowest of the bands seen after extracellular injections (Fig. 3.23). This fiber was reconstructed from serial sections cut at 100 μm in the frontal plane. The branches of the fiber extend through five sections (500 μm) in the rostrocaudal plane. In this plane the central nucleus of this bat measures approximately 1.3 mm, so that the fiber shown here extends roughly one-third of that distance. Thus the axon arborizes in a sheet through a large part of the central nucleus. (D. Raczkowski, E. Covey, J. Casseday, unpublished)

whether the bands from one source interdigitate or overlap with those from another. At present we have only partial answers to the question. Figure 3.23 contrasts direct projections from AVCN with indirect projections from AVCN via INLL. Both project to the ventromedial inferior colliculus in a banded pattern. The bands appear to overlap in some regions of the colliculus, suggesting that they interdigitate within some isofrequency contours. Other evidence from studies of the cat's inferior colliculus, shows that the projections from one LSO interdigitate with those from the other LSO (Shneiderman and Henkel 1987). It seems likely that such interdigitation of ascending projections in a contour is a general principle.

These observations are clues that within the inferior colliculus, a unit of organization consists of a sheet of disc-shaped cells and their associated fibers that together form an isofrequency contour. Each contour receives specific inputs that run parallel to the planar organization of the dendritic ar-

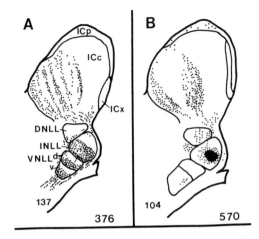

Fig. 3.23 A, B. Drawing of anterograde transport of [³H]-leucine to show convergence at the inferior colliculus of direct and indirect pathways from AVCN. *A* Transport to the nuclei of the lateral lemniscus and inferior colliculus from AVCN. *B* Transport to inferior colliculus from an injection (*blackened area*) in a region of the INLL that is the target of AVCN. Note that the target of the indirect projections overlaps with the target of the direct projections. The orientation of the bands of projections is parallel to the orientation of the disc-shaped cells in the this part of the inferior colliculus. (Zook and Casseday 1987). Abbreviations not found in text: *d,v* dorsal and ventral divisions of VNLL; *ICc,p,x* inferior colliculus, central, pericentral, external divisions

borizations. Thus the projections to the colliculus have a banded appearance that mirrors the laminar arrangement of the sheets of disc-shaped cells. Each organizational unit, or isofrequency contour, receives the same set of inputs from the lower nuclei as the other units, but there is further organization within each unit. These characteristics are most clearly illustrated by the topographic parcellation of binaural response properties in the dorsoposterior division and by the differential subset of lower nuclei that project to each of the aural areas.

3.5 Pathways to the Inferior Colliculus: Conclusions

Three major ideas emerge from the anatomical studies of pathways to the inferior colliculus. These ideas are the source of current investigation. First, there is a major reorganization of the ascending pathways at the inferior colliculus. Second, this reorganization of pathways is reiterated within each isofrequency contour in the inferior colliculus. Third, the AVCN system is further reorganized within each contour and is reflected in the segregation of monaural neurons and different types of binaural neurons within isofrequency contours.

69

An important question to answer in future experiments is whether these pathways converge at the level of single cells. The answer will be important for determining whether or not the monaural pathways play a role in sound localization. For example, does the direct monaural pathway from DCN terminate on the same cells as the binaural pathway from MSO? If so, then sound localization may require more complex integration than simply the timing and intensity processing done at MSO and LSO.

3.6 The Central Acoustic Tract and the Superior Colliculus: Some Speculations on Structures for Spatial Orientation

How does auditory information result in motor action? In order for the bat to respond to the echoes from its sonar emissions there must be some connection between the auditory system and the motor system. The neural computations for determining the location of a sound in space must somehow be translated into motor commands for moving the ears, head and body in relation to the source of the sound. We now describe results that suggest that auditory pathways involving frontal cortex and deep layers of the superior colliculus are important for spatial orientation in bats.

3.6.1 The Central Acoustic Tract

Here we describe an auditory pathway that has its origin in the lower brain stem but bypasses the superior olivary complex and the inferior colliculus. Papez (1929) described an auditory pathway, which he called the "central acoustic tract", that coursed under the inferior colliculus to reach the suprageniculate nucleus, a subdivision of the auditory thalamus. He recognized that it also had a second component of fibers that project to the deep superior colliculus (Cajal 1911). Recently, Kobler et al. (1987) found a pathway in the mustache bat that fits almost exactly Papez's earlier descriptions of projections to the deep superior colliculus and suprageniculate nucleus. Because of this similarity, they adopted the term "central acoustic tract." In addition, they identified the source of the pathway as a group of large multipolar cells just ventral to the nuclei of the lateral lemniscus, which is called here the nucleus of the central acoustic tract (NCAT). The location of NCAT is shown in Fig. 3.24. The fibers of this tract are separate from the main lemniscal system; they travel just medial to the lateral lemniscus, under the inferior colliculus, to end in the deep superior colliculus and the suprageniculate nucleus (Covey et al. 1987; Kobler et al. 1987; Casseday et al. 1988c).

At present we do not know the function of NCAT, but we do know that the pathway that arises from it bypasses the complex interactions that occur in the superior olivary complex, nuclei of the lateral lemniscus and inferior

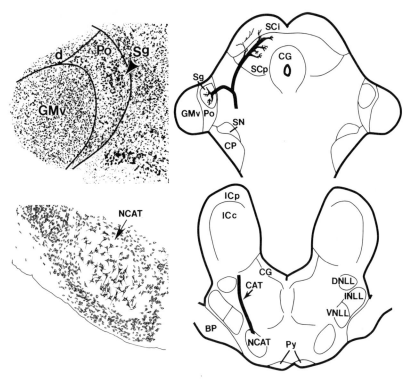

Fig. 3.24. Illustration of components of the central acoustic tract. The origin of the tract is a distinct group of large multipolar cells shown in the drawing at *lower left* (*NCAT*). The thalamic target of these cells is the suprageniculate nucleus (*Sg*), a group of large darkly staining cells located in the posterior thalamic group (*Po*). The origin and terminations of the central acoustic tract (*CAT*) are illustrated in the drawing at the *right*. Note that the pathway is separate from the main lemniscal auditory pathway, i.e., it does not include the central nucleus of inferior colliculus (*ICc*) or the ventral division of the medial geniculate body (*GMv*). Additional abbreviations: *BP* brachium pontis; *CG* central gray; *CP* cerebral peduncle; *d* dorsal division of medial geniculate body; *ICp* pericentral inferior colliculus; *Py* pyramidal body; *SCi,p* superior colliculus, intermediate and deep (proundum) layers; *SN* substantia nigra

colliculus. The potential significance of this pathway is seen by examining further the connections of the superior colliculus, suprageniculate nucleus, and the frontal cortex. Before describing the cortical projections of the suprageniculate nucleus, it is necessary to show why the superior colliculus is an important center for sensory-motor integration.

3.6.2 Superior Colliculus

The connections to the deep layers of the superior colliculus in the mustache bat reveal three points important for the above idea. First, the parts of the superior colliculus that receive auditory input are hypertrophied. These are the

71

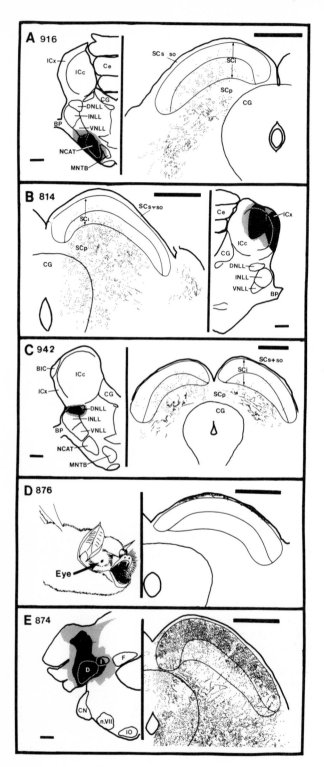

intermediate (SCi) and deep layers (SCp). These layers receive inputs from the NCAT (Fig. 3.25 A), from the inferior colliculus (Fig. 3.25 B) and from the dorsal nucleus of the lateral lemniscus (Fig. 3.25 C). Second, the visual layers are extremely small, as seen by the thin band of projections from the eye (Fig. 3.25 D). The comparison between visual and auditory projections shows that the auditory input is the dominant one of these two sensory inputs. Third, the intermediate and deep layers have connections with motor systems. The input from the deep cerebellar nuclei is shown (Fig. 3.25 E), and there is also output from these layers to motor nuclei for turning the head and moving the ears. Thus the deep and intermediate layers of the superior colliculus are in a position to modulate motor activity, such as ear movements, in response to auditory input.

As yet we do not know the nature of the modulation. However, a clue may be seen by comparison to the visual system. In primates, cells in the deep superior colliculus respond prior to movement of the eyes toward spatial locations of visual or auditory stimuli (Jay and Sparks 1987). These observations suggest that the deep superior colliculus in bats plays an important role in orienting movements of external ears and head toward spatial auditory stimuli.

3.6.3 Connections of Frontal Cortex

To complete our analogy with visual functions, we show how frontal cortex is connected with the subcortical pathways just described. In the mustache bat, the suprageniculate nucleus projects in a rather widespread and diffuse way to all of the auditory cortex (Kobler et al. 1987). But more importantly and more surprisingly, it projects to a second cortical area, anterior to motor cortex. Kobler et al. (1987) showed that this second cortical area can be defined as frontal cortex on the basis of its connections to other thalamic nuclei. In addition to its input from suprageniculate nucleus, the frontal auditory area has reciprocal connections with the auditory cortex. Finally, the frontal auditory field projects to the intermediate and deep layers of the superior colliculus. Figure 3.26 shows this entire set of pathways: the projections from the central acoustic tract to superior colliculus and supragenicu-

Fig. 3.25 A–E. Drawings to show inputs to the superior colliculus of mustache bat. *A* Ipsilateral transport from an injection in the *NCAT*. *B* Ipsilateral transport from an injection in the inferior colliculus. *C* Bilateral transport from *DNLL*. *D* Contralateral transport from the eye to *SCs + so*. *E* Contralateral transport from dentate nucleus of cerebellum. Abbreviations: *BIC* brachium of inferior colliculus; *BP* brachuim pontis; *Ce* cerebellum; *CG* central gray; *CN* cochlear nucleus; *D* dentate nucleus; *DNLL* dorsal nucleus of lateral lemniscus; *F* fastigial nucleus; *I* nucleus interpositus; *IC,c,x* inferior colliculus, central nucleus, external nucleus; *INLL* intermediate nucleus of lateral lemniscus; *IO* inferior olive; *MNTB* medial nucleus of trapezoid body; *n.VII* nucleus of 7th nerve; *NCAT* nucleus of central acoustic tract; *SC,i,p,s + so* superior colliculus, intermediate layer, deep (profundum) layer, superficial gray layers plus stratum opticum; *VNLL* ventral nucleus of lateral lemniscus. Calibration marks: 500 μm. (Modified from Covey et al. 1987)

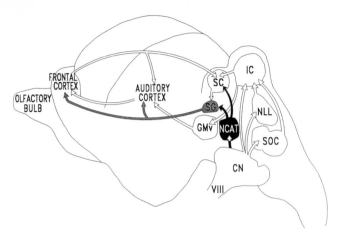

Fig. 3.26. Schematic drawings to show the central acoustic tract, auditory projections to frontal cortex, and related pathways. The central acoustic tract, shown in *black,* terminates in the suprageniculate nucleus and superior colliculus (*SC*). The suprageniculate nucleus (*SG*) projects to auditory cortex (*stippling*), in parallel to pathways from the medial geniculate body. In addition SG projects to frontal cortex (*stippling*). Auditory cortex and frontal cortex have reciprocal connections. Frontal cortex projects to the deep and intermediate layers of the SC. This loop from SC to SG to frontal cortex and back to SC may play an important role in guiding the bat's movements in space. (Casseday et al. 1989)

late nucleus, the cortical extension of this pathway from suprageniculate nucleus to auditory cortex and frontal cortex, and the projection from frontal cortex to the superior colliculus.

Because of its connections with the superior colliculus, Kobler et al. (1987) suggested that the auditory field in frontal cortex plays a role in behavior guided by auditory information, that is, in the acoustic orientation of bats. As mentioned previously, the superior colliculus is a key structure for controlling movements of the head and ears toward external stimuli. In primates, for example, cells in the superior colliculus not only discharge prior to eye movements, but they also control eye movements toward visual or auditory stimuli (Jay and Sparks 1987). The frontal cortex has a number of roles in cognition, and one role that should be common to all mammals is guiding goal-directed behaviors. For example, in primates a specific area of frontal cortex, the frontal eye fields, is important for initiating eye movements toward visual stimuli (Schiller et al. 1980; Goldberg and Bushnell 1981; Bruce and Goldberg 1985). This function is accomplished in concert with the superior colliculus, which receives input from the frontal eye fields (Goldman and Nauta 1976; Kunzle et al. 1976; Illing and Graybiel 1985).

One can see an analogy between vision in primates and echolocation in bats. In somewhat the same way that a primate scans the visual world with eye movements, the bat scans the environment with sound and ear movements to locate objects in space. There is a parallel in the anatomical substrates for these functions: where the primate has visual connections to superior colliculus and frontal cortex, the bat has auditory connections. We suggest that the frontal cortex influences, in a way as yet unknown, motor commands of the superior colliculus to move the ears, head, and body in relation to the source of a sound in space.

3.6.4 Does the Frontal Cortex Play a Role in Spatial Memory?

Spatial orientation involves much more than just these motor reflexes. To orient in space, the bat acquires auditory memory for the spatial location of objects. This kind of spatial memory has been demonstrated experimentally by Neuweiler and Möhres (1967). They trained bats (*Megaderma lyra*) to fly through a grid constructed of wires so closely spaced that the bats had to fold their wings to fly through a gap. Each bat acquired a position habit in that it almost always flew through the same segment of the grid. To test for spatial memory, the wires were replaced by light beams that activated photocells, so that the experimenters could monitor whether the bats still flew in the same location and whether or not the bats continued to fold their wings to "avoid" the obstacle that was no longer present. The light beam was, of course, invisible to the bats. On the following few days of trials, the bats continued to fold their wings and continued to fly through the same location, as if the wires were still present. These bats had obviously developed memory for the spatial objects in their environment. The authors concluded that bats do not rely exclusively on sensory data from echolocation to orient their flight in familiar surroundings, but instead use spatial memory. Consistent with the known cognitive functions of frontal cortex (Crowne 1983), as well as with the results on the mustache bat, is the suggestion that frontal cortex of bats is instrumental in forming spatial memories from auditory cues.

Chapter 4

Physiological Properties of Auditory Neurons

4.1 Introduction

In this chapter we discuss the physiological characteristics of the brainstem auditory nuclei with regard to three major challenges bats face: 1) The determination of target characteristics from the temporal and spectral properties of echoes; 2) the evaluation of target range coded by the temporal interval between an emitted pulse and echo; and 3) the localization of a sound in space as computed from interaural intensity disparities.

We begin with a brief consideration of the auditory nerve. We then focus on the processing and progressive signal transformations that occur within the AVCN and its parallel pathways innervating the LSO, MSO, VNLL and the inferior colliculus. These are the nuclei that have been studied physiologically and they represent many of the principal brainstem nuclei in the primary auditory system. We assume that the underlying connectional basis for sound localization resides, in large part, in the binaural pathways. We further assume that some processes, such as the encoding of target characteristics and range, are the primary responsibility of monaural pathways, because we can see no reason why the encoding of these features should require interaction of the two ears, although this does not preclude the possibility that such information is also conveyed in binaural pathways.

4.2 Auditory Nerve

The echoes of insects are complex. The echo CF component that stimulates the ears of horseshoe and mustache bats contains pronounced amplitude and frequency modulations that have periodicities and other features determined by the insect's wingbeat frequency, wing size, and orientation with respect to the bat. The chief clue about how insect echoes are encoded arose from experiments that employed electronically generated modulated signals that mimicked the echoes from a fluttering insect. Suga and Jen (1977) were the first to describe the coding of such modulation patterns in the auditory nerve of the mustache bat. They mimicked insect echoes by modulating the frequency or amplitude of a 60 kHz carrier tone with low frequency sinusoids of 50–100 Hz. We hereafter refer to this low frequency sinusoid as the modulating waveform, to distinguish it from the 60 kHz carrier frequency. This ar-

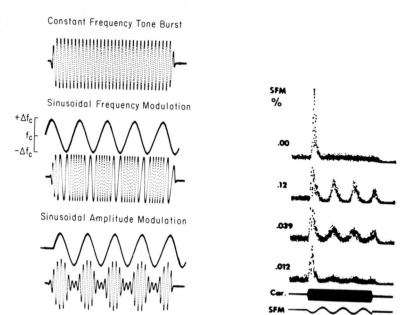

Fig. 4.1 (*Figure to the left*). Illustration of fine structure of a tone burst (*upper record*), a sinusoidally frequency modulated (SFM) burst (*middle record*) and a sinusoidally amplitude modulated (SAM) burst (*lower record*). The tone burst in the upper record is a shaped sine wave, where the frequency, or fine structure, of the signal remains constant. The fine structure of the SFM burst changes sinusoidally in frequency. The modulation waveform is shown above the record of the signal's fine structure. In the SAM burst, the fine structure of the signal is a constant frequency, but the amplitude varies with the sinusoidal modulating waveform

Fig. 4.2 (*Figure to the right*). Peri-stimulus time histograms of the discharges of a filter unit in the auditory nerve of the mustache bat evoked by tone bursts (*uppermost record*) and by SFM signals of varying modulation depths. The modulation depth is expressed as % SFM. The first peak in each histogram is the neuron's discharges to the onset of the stimulus. The subsequent peaks are the discharges that are phase-locked to the SFM waveform. Small peaks are discernable when the modulation depth was only 0.012% of the 61.90 kHz carrier frequency. The modulating frequency was 100 Hz and the signal intensity was 80 dB SPL. (Data from Suga and Jen 1977)

rangement created either sinusoidally amplitude modulated (SAM) or sinusoidally frequency modulated (SFM) signals (Fig. 4.1). With SFM signals the depth of modulation, the amount by which the frequency varies around the carrier, mimics the degree of Doppler shift created by the motion of the wings. The other modulating parameter, the modulation rate, simulates the insect's wingbeat frequency.

Suga and Jen (1977) observed that 60 kHz filter neurons routinely phase-lock to the modulating waveform of an SFM signal, and some do so even when the modulation depth varied by as little as ± 10 Hz around the 60 kHz carrier frequency (Fig. 4.2). Such sensitivity for frequency modulation is truly remarkable. Although primary auditory neurons in other animals also display phase-locking to SFM signals, such high sensitivity occurs only in filter neurons. It seems reasonable to conclude, as did Suga and Jen, that the sharp tuning appears to be a specialization for the fine analysis of frequency modulation patterns imposed on the echo CF component.

Suga and his colleagues (Suga et al. 1975) also observed that filter units discharge in phase with the envelope of amplitude-modulated signals. Phase-locking to amplitude modulations, however, is not a unique property of filter units. Phase-locking to amplitude-modulated signals is seen in the auditory nerve of other mammals (Møller 1976), and it is readily evoked in neurons tuned to any frequency in the bat's audible range. These experiments indicate that the encoding of amplitude modulations probably plays a smaller role than the encoding of frequency modulations in the detection of insect echoes.

4.3 Anteroventral Cochlear Nucleus

Discharge patterns of neurons localized in the AVCN have been described by Feng and Vater (1985) in horseshoe bats. The most common pattern, found in about half of the cells in this region, is a primary-like pattern (Fig. 4.3), a term originally used by Pfeiffer and Kiang (1965) because of the similarity of the discharge pattern of some AVCN cells in the cat to the discharge pattern evoked by tone bursts in auditory nerve fibers. This pattern is characterized by an initially high discharge rate which then adapts to a lower rate that is

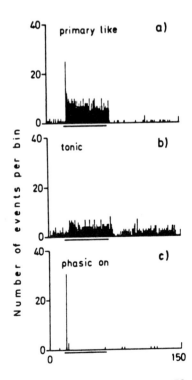

Fig. 4.3 a–c. Peri-stimulus time histograms showing the main types of discharge patterns of neurons in the AVCN of the horseshoe bat. (After Feng and Vater 1985)

maintained for the duration of the tone burst. Other AVCN neurons, however, respond with pure excitation, without the phasic component, or with a pure phasic-on pattern (Fig. 4.3).

It is tempting to suggest that the AVCN neurons discharging with a primary-like pattern are spherical and globular cells as Rhode and his colleagues have shown with intracellular recordings in the cat's cochlear nucleus (Rhode et al. 1983; Rhode 1985; Smith and Rhode 1987). These cell types dominate the AVCN, and in other animals receive calyciform and large bouton synaptic endings from auditory nerve fibers (e.g., Cant and Morest 1984). In the previous chapter we pointed out that these secure synapses are considered to permit a faithful one-to-one transmission from pre-to-postsynaptic cell.

In a study of response properties evoked by modulated signals, Vater (1980, 1982) found that many neurons in the cochlear nucleus of the horseshoe bat synchronize their discharges to the period of SFM and SAM stimuli, as was seen by Suga and Jen in the auditory nerve. Amplitude-modulated signals evoked responses synchronized to the modulating waveform in almost all cochlear nucleus cells, and this was equally apparent in filter neurons and in neurons tuned to other frequencies.

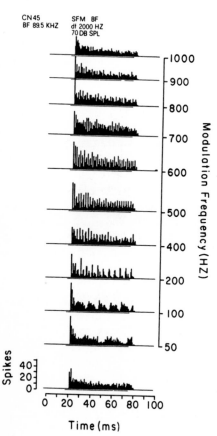

Fig. 4.4. Responses of a neuron in the horseshoe bat's AVCN to SFM stimuli of various modulation frequencies. The modulation frequency that generated each histogram is shown on the right. The modulation depth was ±1000 Hz. The carrier frequency was 89.5 kHz at 70 dB SPL. (Data adapted from Vater 1982)

80

In contrast, the phase-locked responses to SFM stimuli were more variable. Vater (1980) suggested a relationship between the response property and the structure of the synaptic connections within the divisions of the cochlear nucleus. Primary-like units in the AVCN responded best to SFM signals, and their discharges were synchronized at modulation frequencies up to 900 Hz (Fig. 4.4). Some phasic-on neurons in the DCN also displayed phase-locked discharges to SFM but only to low modulation rates up to about 200 Hz. Other neurons that exhibited more complex discharge patterns to tone bursts were unable to follow the modulation waveforms and responded to SFM signals as they did to tone bursts. She found, moreover, that these features are not restricted to filter neurons, but are seen in neurons tuned to most frequencies of the horseshoe bat's hearing range.

The distinguishing characteristic of filter units in the cochlear nucleus is their extreme sensitivity to modulation depth. They are not more likely to phase-lock to modulated signals than are neurons tuned to other frequencies, but rather they are much more sensitive to small frequency changes. In common with the findings of Suga and Jen in the mustache bat, Vater found that filter neurons, and only filter neurons, would phase-lock to modulation depths as low as ± 20 Hz around an 80,000 Hz carrier frequency.

4.4 Medial Superior Olive

The MSO is a binaural nucleus that receives direct innervation from the AVCN bilaterally. Before describing the properties of MSO neurons, we introduce nomenclature used with binaural cells. The type of cell is indicated with two letters; the first refers to the input from the contralateral, and the second refers to the input from the ipsilateral ear. The input from each side is indicated with an "E" for excitation, an "I" for inhibition or an "O" for no effect. Thus an E-I cell receives excitation from the contralateral ear and inhibition from the ipsilateral ear, whereas an E-E neuron receives excitation from both ears. Monaural neurons are designated as E-O and receive innervation from only one ear. Examples of each type are shown in Fig. 4.5.

Harnischfeger et al. (1985) recorded from neurons in the superior olivary complex of *Molossus ater* while presenting acoustic stimuli through earphones so that interaural intensity and timing parameters could be varied independently. They introduced delays to one ear while holding the intensities presented to the two ears constant, and in other situations they changed the intensities at the two ears while keeping interaural timing constant. They found that stimulation of the excitatory ear evoked a tonic discharge pattern in 75% of MSO neurons. The aural types of most MSO neurons were E-E or E-I, and occurred in almost equal numbers, but a small number of monaural neurons were also found.

One of the most significant findings was that a few MSO neurons were sensitive to interaural timing differences around ± 50 μs, and several others

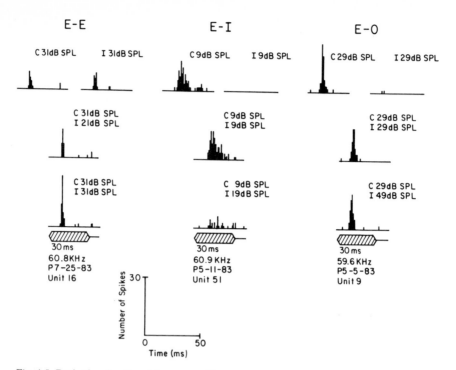

Fig. 4.5. Peri-stimulus time histograms illustrating the major types of binaural neurons in the auditory system. A binaural neuron receiving excitatory input from both ears, E-E cell, is on far *left;* excitatory-inhibitory, E-I cell, is in *middle panel;* and a monaural, E-O cell, is on *far right.* Stimulus to contralateral ear is indicated by *C,* and stimulus to ipsilateral ear is indicated by *I.* All neurons were recorded from mustache bat's inferior colliculus

were sensitive to interaural timing differences of ± 500 μs. The changes in discharge with small interaural timing differences in a time-sensitive MSO neuron are shown in the lower panel of Fig. 4.6. The temporal feature of importance for these cells is the onset disparity, since the cells are insensitive to the phase disparities of the high frequency sinusoids. The sensitivity to particular interaural time differences, however, is only evident with specific interaural intensity differences. The changes in the discharges of the cell in Fig. 4.6, for example, were obtained with a constant interaural disparity of 7 dB, where the intensity at the excitatory (contralateral) ear was 7 dB greater than the intensity at the inhibitory (ipsilateral) ear. Moreover, all time-sensitive cells were also sensitive to changes of interaural intensity disparities, in that the discharge rate fell progressively as the intensity at the inhibitory ear was increased.

As we pointed out in Chapter 1, interaural timing disparities in the range of ± 50 μs are created by sounds located at different azimuthal positions around the bat's head. The ability of bats to utilize time disparities for sound localization has not previously been given much credence, but neurons exhibiting sensitivities to these small temporal disparities could be important

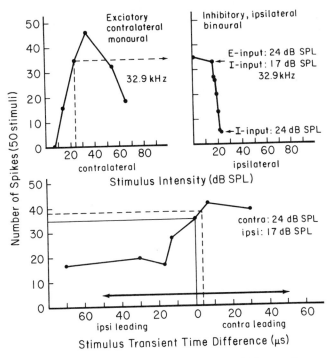

Fig. 4.6. Response properties of a time sensitive E-I neuron from MSO of *M. ater. Bottom panel* Changes in discharge rate with small changes in interaural time differences in the microsecond range. The *solid lines* show the discharge rate at a time disparity of 0 µs, and the *broken lines* show the discharge rate at a time disparity of +3 µs, contralateral (excitatory) ear leading. The interaural intensity disparity was 7 dB, with the excitatory ear louder. *Upper left panel* Discharge rate as a function of increasing sound intensity at excitatory ear. No sound was presented to inhibitory ear. *Broken lines* show discharge rate evoked by monaural stimulus of 24 dB SPL to excitatory ear. *Upper right panel* Decline in discharge rate as intensity to inhibitory ear is increased. The sound intensity at excitatory ear was held constant at 24 dB SPL, and interaural time disparity was 0 µs. *Arrow at top* shows discharge rate when intensity at excitatory ear was 24 dB SPL and intensity at inhibitory ear was 17 dB SPL. *Arrow at bottom* shows discharge rate when intensity at excitatory ear was 24 dB SPL and intensity at inhibitory ear was also 24 dB SPL. All tone bursts were 32.9 kHz. (Data adapted from Harnischfeger et al. 1985)

for encoding sound location. We discuss the possible role of time-sensitive units for encoding the azimuth of a sound following a description of the binaural properties of LSO neurons.

4.5 Lateral Superior Olive

In cats the LSO is a homogeneous nucleus in terms of the discharge patterns of its neurons (Tsuchitani 1982) and binaural type, which are predominantly

I-E (Boudreau and Tsuchitani 1968; Tsuchitani 1977). Almost all neurons receive excitation from the ipsilateral ear and respond with a tonic response pattern. Inhibitory inputs originate from the contralateral ear, via the MNTB. Harnischfeger et al. (1985) recorded from ten LSO cells in *M. ater,* all of which had properties similar to those reported for the cat's LSO. In contrast to the MSO neurons, LSO neurons in *M. ater* are not sensitive to small time differences. Rather they are sensitive predominantly to interaural intensity differences.

This finding suggests a fundamental difference in mechanism between the LSO and MSO in *M. ater.* MSO neurons in the auditory system of *M. ater* appear to function as coincidence detectors, and thus are very sensitive to the difference in the arrival times of discharges from the two ears. LSO neurons in *M. ater,* on the other hand, are apparently much less sensitive to the precise temporal interval between the arrival of inputs from the two ears. Rather the LSO neurons seem to compare the "amount" of activity arriving from the excitatory ear with the "amount" of activity arriving from the inhibitory ear over time periods that are much longer than those upon which MSO neurons operate. The "amount" of activity arriving from each ear could be the difference in the discharge rate of the excitatory and inhibitory neurons, a difference in the number of fibers excited by each ear, or a combination of the two factors.

A functional difference between the LSO and MSO is consistent with the ideas that arose from studies of other mammals. In cats and other large terrestrial animals, the LSO is thought to process only intensity disparities whereas the MSO processes interaural time disparities (Erulkar 1972; Masterton and Diamond 1973; Masterton et al. 1975). But in these animals there is a pronounced difference in frequency representation in the two regions. In the LSO high frequencies are overrepresented (Tsuchitani and Boudreau 1966; Guinan et al. 1972b), and in the MSO low frequencies are overrepresented (Goldberg and Brown 1968; Guinan et al. 1972b).

In the early studies of the cat's LSO, tests were made only for responses to intensity disparities and not for sensitivity to temporal disparities. Two recent studies of the cat's LSO (Caird and Klinke 1983; Yin et al. 1986), however, show that LSO neurons also process interaural time disparities in a fashion qualitatively similar to that reported for MSO neurons. Of particular interest is that Yin and his colleagues (Yin et al. 1986) report that high frequency cells in the cat's LSO, which code for interaural intensity disparities, are also sensitive to interaural time disparities of amplitude-modulated waveforms. The LSO neurons changed their discharge rates with interaural phase differences of the amplitude-modulated envelope but were insensitive to the phase disparities of the high carrier frequency.

In view of the meager information concerning the temporal sensitivity of LSO neurons in any mammal, it is difficult to identify functional differences between neurons in the MSO and LSO with any degree of confidence. Moreover, there is a paucity of information about response properties of MSO neurons, and to further complicate the issue, there is debate about whether the studies of the cat's MSO actually sampled MSO neurons or neurons from adjacent nuclei. Perhaps it is safest at the moment to say that defining the distinctions between LSO and MSO remains as one of the important questions in auditory physiology.

4.6 How Do Time-Sensitive E-I Units Code for Azimuth?

The work of Harnischfeger et al. (1985) suggests that the E-I neurons in the LSO are sensitive to interaural intensity disparities and are relatively insensitive to time disparities. It seems clear that these neurons code for the intensity differences, at the two ears, that occur with sounds at different azimuthal positions. Time-sensitive MSO E-I neurons, however, are very sensitive to both interaural intensity and time disparities. One can ask whether these neurons code for both time and intensity disparities, and if so specifically how does this work. There are two functional interpretations that can be given to the properties of time-sensitive neurons. The first, which we shall call the complementary interpretation, is that both the disparities in time and intensity are coded to assess the azimuth of a sound. Since changes in the horizontal location of a sound simultaneously produce both a change in interaural interaural time and intensity (Fig. 4.7), the neuron's discharge rate also changes as a consequence of its sensitivity to the changes in time and intensity disparities. Thus the neuron's responses to the two cues should complement one another to produce a larger change in discharge rate than would be produced by either cue by itself.

   The second interpretation is that the only relevant cue is the intensity disparity. The reasoning is that the *acoustic cue* upon which the nervous system

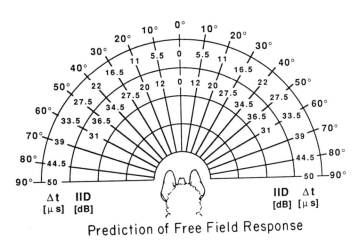

Prediction of Free Field Response

Fig. 4.7. Interaural time (*Δt*) and intensity (*IID*) disparities that occur at various azimuthal positions around the head of *M. ater* with 32-kHz tones. IID values were measured from N_1 responses, and *Δt* values were calculated. (After Harnischfeger et al. 1985)

works is the intensity disparity at the two ears, but *binaural processing* operates upon differences in the arrival time of the discharges from the two sides of the brain. The differences are produced by trading time for intensity. The acoustic temporal disparity, the disparity in the arrival time of the sounds that reach the two ears, is so small that it is inconsequential in this scheme. The way this could work is explained below.

Binaural neurons having high sensitivities for disparities of both inter-aural time and intensity exhibit time-intensity trading, a process that has also been demonstrated in binaural neurons of other mammals (e.g., Hall 1965; Moushegian et al. 1964, 1967; Caird and Klinke 1983; Yin et al. 1985) and in human psychophysical studies (Moushegian and Jeffress 1959; David et al. 1959; Deatherage and Hirsh 1959; Witworth and Jeffress 1961; Hafter 1984). A trade of time for intensity requires that the nervous system convert the intensity disparity into a differential firing latency from the two ears, and that within the nervous system, time-sensitive binaural cells function in a manner similar to coincidence detectors. The factor that determines whether or not a binaural cell will discharge is the timing of the inputs from the inhibitory and excitatory ears (Jeffress 1948; Deatherage and Hirsh 1959; Yin et al. 1985). When the excitatory input arrives earlier than the inhibitory input the cell discharges, but if the excitatory and inhibitory inputs are coincident, or if the inhibitory input arrives earlier than the excitatory input, the cell fails to discharge. Since firing latency shortens as signal intensity is increased, making the sound louder at one ear should effectively create a shortening of the discharge latency arriving from that ear. That latency decrease can be offset by advancing in time the signal to the opposite ear by a comparable amount thereby re-establishing the original binaural temporal pattern arriving at the neuron. Time-intensity trading ratios ranging from about 8–50 $\mu s/dB$ were seen in MSO and inferior colliculus neurons in *M. ater* (Harnischfeger et al. 1985), and comparable values were recently found in binaural cells of the inferior colliculus of Mexican free-tailed bats (Pollak 1988).

It is instructive to consider the impact of the changes in interaural time and intensity differences on the discharge rate of the neuron in Fig. 4.6. Consider first the firing rate of a sound that is 7 dB louder in the contralateral ear than the ipsilateral ear, the intensity disparity used to test this neuron's time sensitivity. According to Fig. 4.7, this intensity disparity will be generated when a sound is located about 6 degrees in the contralateral sound field, and a sound at this location will also generate a time disparity of about 3 μs, where the sound arrives earlier at the contralateral ear. The dashed line in the lower panel of Fig. 4.6 shows that the neuron discharged about 38 spikes per 50 stimuli when binaurally stimulated with these disparities. This is a slightly higher rate than the 35 spikes per 50 stimuli evoked by the excitatory signal presented alone at 24 dB (dashed lines in upper left panel of Fig. 4.6). The difference in spike rate between these two conditions is almost certainly a consequence of random variation in the discharge function. The critical question is whether a small change in temporal disparity, as occurs when the sound location is shifted to 0 degrees azimuth, can be coded by this neuron? The solid line in the lower panel of Fig. 4.6 shows that when the time disparity was changed to 0 μs, which occurs when the sound is located at 0 degrees azimuth, the discharge rate changed only slightly to about 35 spikes per 50 stimuli. However, and this is very important, at 0 degrees azimuth the intensity disparity becomes 0 dB. When the intensity at both ears was equal (both signals were 24 dB) and the temporal disparity was 0 μs, the firing rate dropped markedly to about 3–4 spikes per 50 stimuli, as shown in the upper right panel of Fig. 4.6. These properties show that this neuron must have a high degree of spatial selectivity, since the discharge rate changes dramatically for disparities that are created at locations between 0–6 degrees along the horizontal axis. But clearly, the 7-dB change in intensity disparity that occurs along these locations causes a major decline in discharge rate whereas the change in discharge rate resulting only from an interaural timing difference of 3 μs is within the range of random variations.

The above analysis favors the second interpretation given above, that it is only the interaural intensity disparities that determine the spatial properties of this neuron. But what then is the significance of the neuron's very high temporal sensitivity? The most likely explanation is that the temporal sensitivity is, in reality, an expression of extreme sensitivity to interaural intensity disparities, which are converted into time disparities in the brain. Let us consider how this might work. As best as can be judged, the time-intensity trading ratio of this neuron is on the order of about 8 μs/dB: every intensity change of one decibel causes an 8 μs latency change. When the intensity to the contralateral ear is held constant, the discharge latency from that ear is also constant, but every dB increase to the inhibitory ear causes the inhibitory inputs to arrive 8 μs earlier at the MSO. Thus a 7-dB increase at the inhibitory ear should cause the inhibitory input to advance in arrival time at the MSO neuron by about 56 μs. The point is that when sound location changes from 6 degrees to 0 degrees, the ears "see" a change in timing disparity of only 3 μs, but the MSO neuron "sees" an amplified interaural timing change of 50–60 μs because the nervous system trades time for intensity. Furthermore, due to its sensitivity for these time disparities, as illustrated in Fig. 4.6, the amplified arrival time difference from the excitatory and inhibitory ears causes a major reduction in the neuron's discharge rate.

4.7 Inferior Colliculus

4.7.1 Temporal and Spectral Coding Evoked by FM Signals Related to Target Ranging

One of the consistent transformations that occur between the lower brainstem auditory nuclei and the inferior colliculus is a change in discharge pattern evoked by tone bursts, from varieties of predominantly tonic patterns in the lower regions to predominantly phasic-on patterns in the inferior colliculus (e.g., Suga 1973; Neuweiler and Vater 1977; Pollak and Schuller 1981; Vater 1982; Harnischfeger et al. 1985). In the section below we consider the monaural pathway from the AVCN leading to the columnar VNLL and the projection from AVCN to the inferior colliculus. We suggest a relationship between the VNLL and a variety of phasic-on inferior colliculus neurons and describe the presumptive role of these collicular neurons for temporal and spectral coding. Evidence for temporal coding by bats comes from behavioral studies concerned with target ranging (Simmons 1973, 1979). These experiments showed that the cue for range measurement is the time interval between the reception of the FM components of the pulse and echo. Bats are able to discriminate between the time of occurrence of an emitted pulse and the time of occurrence of a returning echo, with a precision such that they can detect changes as small as 75 μs, perhaps even as small as 1–2 μs. These values correspond to accuracies of target range detection of about 10.0 mm and less than 0.5 mm respectively.

A population of cells in the inferior colliculus of Mexican free-tailed bats have response characteristics well suited for target ranging. These cells have a phasic-on discharge pattern and discharge latencies that vary only slightly when the same signal is presented repeatedly (Pollak et al. 1977; Pollak 1980; Bodenhamer et al. 1979; Bodenhamer and Pollak 1981). Neurons having these properties are called constant latency neurons, and are distinguished by four key features: 1) a phasic-on response pattern where the cell discharges at most two spikes to an FM stimulus; 2) a reliable discharge probability to every or nearly every stimulus presentation for intensities 2–6 dB above threshold or greater; 3) a highly consistent initial discharge latency over a wide intensity range; and 4) a relatively constant response latency where the average latency change is 0.75 ms over a 30-dB intensity range. Each of these features is illustrated by the constant latency neuron in Fig. 4.8.

When driven by a brief FM signal, constant latency neurons respond to only one frequency, or a very narrow band of frequencies, in the FM signal and not to the signal onset (Fig. 4.9). When such a brief FM burst is presented repeatedly, the neuron's discharges lock onto the same frequency component, and the latencies fall into tight registration. The frequency of an FM burst to which the discharges are locked is called the cell's excitatory frequency. As shown in Fig. 4.9, when the duration of an FM signal is changed, but the starting and ending frequencies are kept constant, the excitatory frequency will shift in time as the duration of the FM signal is increased. Discharge latency shifts in accordance with the frequency to which the cell is locked (Bodenhamer et al. 1979; Bodenhamer and Pollak 1981).

The implications of these features for the processing of biosonar signals are considerable. Constant latency neurons are highly selective for both temporal and spectral attributes of the biosonar signals and are well suited for encoding both target distance and the spectral composition of the echo, which is important for target identification.

The fact that constant latency neurons record the temporal event for a particular frequency in an FM pulse means that these neurons can serve as precise time markers. These neurons should encode the time of occurrence of

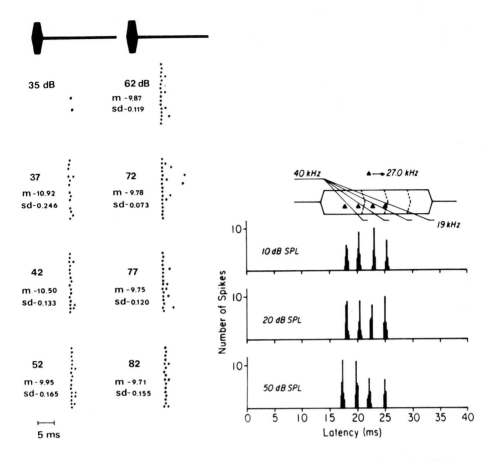

Fig. 4.8 (Figure to the left). Firing pattern of a constant latency neuron from the Mexican free-tailed bat's inferior colliculus. Each *dot column* shows discharges to 16 stimulus presentations at intensities shown at *left*. At 35 dB the neuron discharged two action potentials to 16 presentations of the FM burst, whereas at 37 dB, 14 spikes were evoked. Note tight registration of initial discharges that are evident even at 37 dB. Mean initial discharge latency (*m*) and standard deviation (*SD*) of each dot column are shown at *left*. Stimuli were 3.4 ms FM pulses that swept downward from 50–25 kHz. (Pollak et al. 1977)

Fig. 4.9 (Figure to the right). Responses of a constant latency cell to four durations (4.0, 6.0, 8.0, and 10.0 ms) of an FM burst that swept downward from 40 kHz to 19 kHz. Each stimulus was presented 25 times at a given duration and intensity. The peri-stimulus time histograms have been aligned directly beneath the FM sweeps, and demonstrate that the discharges of this cell were synchronized to the 27.0 kHz component in the FM signal (indicated by *filled triangle*). The first column of histograms was generated when the duration of the FM sweep was 4.0 ms, at intensities of 10, 20, and 50 dB SPL. The second column of histograms was generated with 6.0 ms FM sweeps, the third column with 8.0 ms sweeps and the fourth column with 10.0 ms FM sweeps. Note that the discharges remained synchronized to the 27 kHz component of the sweep at all intensities. (Bodenhamer and Pollak 1981)

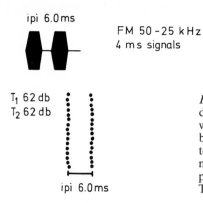

ipi 6.0 ms

FM 50 – 25 kHz
4 ms signals

T₁ 62 db
T₂ 62 db

ipi 6.0 ms

Fig. 4.10. Responses of a constant latency unit to two downward FM bursts, illustrating the precision with which these neurons record the temporal interval between two pulses. The temporal, or interpulse interval (*ipi*) was 6.0 ms, and the intensity of both signals was 62 dB SPL. Each *dot pair* represent action potentials evoked by the first and second FM burst. The bursts were presented repetitively

a particular frequency component in the emitted pulse and the time of occurrence of the same frequency in the returning echo. The period between the occurrences of a frequency in the pulse and echo is a function of the pulse-echo delay, and hence the target distance. As shown in Fig. 4.10, constant latency neurons encode the temporal separation between two, equally intense FM signals with remarkable precision.

The timing precision illustrated by the constant latency neurons in Fig. 4.10 is a necessary but not a sufficient demonstration to warrant a conclusion that these neurons encode range information. The two FM bursts were of equal intensity and were presented at only one time interval. Since target range must be evaluated at a variety of distances with a wide variety of echo intensities, it is important to show that discharge characteristics, similar to those in Fig. 4.10, can be elicited with signals simulating the various pulse-echo combinations that would occur as a bat homes in on its target.

Two features of the constant latency neurons are important in this regard. The first is that the neuron's excitatory frequency is nearly invariant with intensity. Consequently, the discharges of these cells remain synchronized to the same, or nearly the same, excitatory frequency over a wide range of intensities (Fig. 4.9) (Bodenhamer et al. 1979; Bodenhamer and Pollak 1981). Thus the neuron should discharge to the same frequency of both the loud emitted pulse and fainter echo, thereby preserving the timing accuracy. The second feature concerns the fast recovery times of constant latency neurons, where the presentation of a loud initial signal has almost no influence on the response to a much fainter second signal following shortly thereafter (Pollak et al. 1977). These characteristics allow the constant latency neurons to encode the temporal period between an emitted pulse and echo under conditions encountered during echolocation (Fig. 4.11).

The precision with which constant latency neurons encode the temporal interval between two FM bursts, however, is somewhat influenced by the relative intensities of the two signals; the discharge latencies evoked by louder signals are slightly shorter than the latencies evoked by fainter signals. It follows then that a loud pulse followed by a fainter echo will, in any given constant latency neuron, evoke two action potentials, but the discharge inter-

90

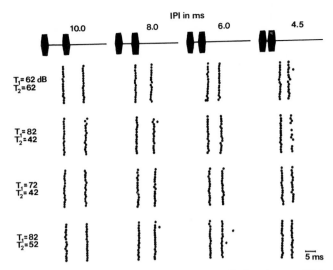

Fig. 4.11. Recovery of responsiveness in a constant latency neuron. All signals were FM bursts that swept downward from 50–25 kHz. The intensities of the first and second signals are shown as T_1 and T_2 respectively. Note the fast recovery and the precision with which the response intervals follow the inter-pulse-interval (IPI), even when the first FM (simulated emitted pulse) was 30 dB more intense than the second FM signal (simulated echo). Each *pair of dot columns* was generated by 16 repetitions of simulated pulse-echo combination

val will be slightly longer than the temporal interval of the signals. In this context it is important to recall that one feature of constant latency neurons is that their discharge latencies increase on the average by only 0.75 ms over a 30-dB intensity range (Pollak et al. 1977). Thus, if the bat emits a pulse where the intensity reaching the cochlea is 82 dB and receives an echo 5 ms later at an intensity of 52 dB, the average constant latency neuron should respond to each signal with discharges separated by 5.75 ms. The neuron would encode a distance of 98 cm instead of the true distance of 86 cm, an error of 12 cm. The error depends only upon the pulse-echo intensity difference; the smaller the intensity difference the less the timing error. The important point is that the possible range of errors is sharply limited because of the small latency variation with intensity.

Such errors in an individual neuron are not difficult to reconcile with the ranging precision required of bats. The timing error depends only upon the pulse-echo intensity difference, and since the bat encounters these errors throughout life, it must somehow learn to make appropriate corrections. Presumably, the bat could associate each pulse-echo intensity difference with a given error, and thereby correct for it.

Although constant latency neurons have been studied in detail only in the inferior colliculus of Mexican free-tailed bats, neurons having similar features have been found in a variety of other species and seem to be common among bats (Suga 1970; Vater and Schlegel 1979; Pollak and Bodenhamer 1971; Pollak and Schuller 1981).

91

It is tempting indeed to suggest that the population of constant latency neurons is the neuronal population in the midbrain which receives innervation predominantly from the columnar division of the VNLL. This explanation would certainly be parsimonious because the sequence of calyciform synapses across rows of precisely oriented isofrequency neurons is an architectural arrangement well suited for preserving information. Nevertheless, we also emphasize the speculative nature of the proposal, since it is by no means certain that neurons classified as constant latency neurons are monaural, or that they are, in fact, innervated by projections from the VNLL. Clarifying the relation between the constant latency neurons and the columnar VNLL is another challenge for future studies.

4.7.2 Further Coding of Target Range

A report by Feng et al. (1978) suggests that range is coded in higher centers by a place code. These investigators recorded from single units in the big brown bat's intercollicular nucleus, a poorly defined region of the midbrain lying between the inferior and superior colliculi. They found a population of neurons that did not respond to tonal or FM stimuli, but did respond vigorously to a pair of FM bursts with particular "pulse-echo" delays. The interpretation they gave to these results was that target range may be encoded by a neural place mechanism incorporating different delay-tuned neurons in the central neurons system, presumably at levels above the inferior colliculus.

The preferred pulse-echo delay times exhibited by intercollicular neurons were quite broad. The width of the delay tuning was typically 5–8 ms which would not satisfy the ranging requirements of echolocating bats. An alternate interpretation is that neurons selective for a range of pulse-echo delays may be important for eliminating clutter produced by echoes from closer and more distant objects and thus may allow the bat to focus upon one, primary target of interest.

The report by Feng et al. (1978) was also important because it suggested that neurons at even higher levels of the auditory system might have delay-tuning properties much sharper than the intercollicular neurons. This possibility prompted Suga and his colleagues to look for neurons with these properties in the cortex of the mustache bat, where they discovered an elegant topographic mapping of target range by sharply tuned delay-sensitive neurons (Suga and O'Neill 1979; 1986; Suga 1984; 1980).

4.7.3 Spectral Coding for Target Attributes

While the constant latency neurons represent the extreme in fidelity of discharge latency to simulated orientation sounds, other collicular neurons also respond to brief FM signals with a phasic-on pattern, discharging from 1–3 action potentials depending upon the intensity of the signal. These neurons

92

also have a clearly defined excitatory frequency, a feature in common with the constant latency neurons.

An additional feature of some interest is that constant latency and other collicular neurons are much more selective for frequency when driven with brief FM stimuli than when driven with tone bursts. Classically, frequency selectivity of auditory neurons is described in terms of tuning curve sharpness, as determined with tone burst stimulation. Tuning curves of auditory neurons in FM bats have Q_{10dB} values ranging from about 4–15. However, it is biologically more meaningful to describe the selectivity of constant latency neurons in terms of their responses to FM signals, since it is from FM signals that bats extract much or all of their echolocation information. When neurons are driven with tone bursts, the range of frequencies capable of evoking discharges often becomes quite large at higher intensities (i.e., the tuning curve broadens). On the other hand, it appears that a frequency band of 0.5–2.5 kHz is responsible for triggering the majority of neurons with FM signals, and this bandwidth remains stable with intensity (Bodenhamer et al. 1979; Bodenhamer and Pollak 1981). As a consequence, the widths of the frequency bands of FM stimuli that evoked responses from these neurons are much narrower than are their tuning curves obtained with tone bursts. Viewed in this way, the range of frequencies to which these neurons respond is comparable to the range of frequencies that excite the filter neurons of long CF/FM bats (Fig. 4.12). Apparently determining frequency selectivity with tone bursts is not exactly comparable to obtaining excitatory frequencies from discharge latencies to FM signals.

Recall that a neuron's best frequency is thought to be a reflection of the place along the cochlear partition from which that neuron receives its primary innervation. There is a close and consistent relationship between a neuron's best frequency to tone burst stimuli and its excitatory frequency obtained with brief FM signals, with each neuron's excitatory frequency slightly exceeding its best frequency by 2–4 kHz. It should be noted, however, that the comparison is not between two corresponding parameters. The best frequency is defined for only one intensity (i.e., the lowest capable of discharging the cell) whereas the excitatory frequency can be determined for any suprathreshold intensity. Since the best and excitatory frequencies are closely correlated, the orderly representation of best frequencies is similar to the orderly progression of excitatory frequencies along the dorsoventral axis of the Mexican free-tailed bat's inferior colliculus, where units with low excitatory frequencies are located dorsally and those with high excitatory frequencies are ventral (Fig. 4.13) (Bodenhamer and Pollak 1981).

The place coding of excitatory frequencies provides a mechanism for ensembles of collicular neurons to recreate the biosonar signal in time, frequency, and amplitude across the tonotopically organized neural tissue of the inferior colliculus (Fig. 4.11). This is important because the echoes bats receive often resemble filtered replicas of the emitted pulses, and the way in which a target "filters" the emitted signal, i.e., modifies its power spectrum, is an important clue about its structure. Each frequency of an FM burst occurs sequentially in time, and there is further spread in time due to the differential

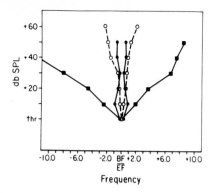

Fig. 4.12. Figure showing the relative frequency selectivity of a typical neuron in the Mexican free-tailed bat's inferior colliculus (*squares*) compared to the frequency selectivity of a filter neuron and the frequency selectivity of a neuron determined with FM bursts. This neuron's best frequency (BF) was 52.0 kHz and the Q_{10dB} of its tuning curve was 11. *Open circles* show the tuning curve determined with tone bursts of a typical filter neuron that was recorded from the mustache bat's inferior colliculus. The filter neuron's best frequency was 61.6 kHz and the Q_{10dB} of its tuning curve was 147. *Filled circles* show the frequency selectivity of a constant latency neuron from the Mexican free-tailed bat's inferior colliculus determined from its responses to FM stimulation. This last curve (*filled circles*) was obtained by determining the cell's excitatory frequency (EF) from its discharge latencies to four durations of an FM sweep at various intensity levels. The *filled circles* indicate the range of frequency components that triggered the cell as determined from the discharges evoked by the signals having different durations. The "tuning curve" is centered over the mean EF. Comparison of the tone burst and FM responses demonstrates that FM signals sharpened the frequency selectivity of cells in the Mexican free-tailed bat's inferior colliculus. (Bodenhamer and Pollak 1981)

it takes the traveling wave to reach more apical regions of the cochlea. Thus the neurons sychronized to high frequencies should be the first to discharge, followed by units having progressively lower excitatory frequencies. If each frequency in an FM signal excites a population of neurons in the corresponding isofrequency sheet of collicular cells, then the intensity of that particular frequency should be coded by the number of neurons responding in that frequency contour, by their discharge rates and their latencies.

One can imagine that notches in certain spectral regions of the echo could create similar notches, reduced firing rates or a reduced number of discharging cells, in the collicular contours that represent the attenuated frequencies. Such notches can arise when a complex signal, such as an FM burst, is mixed with a time-delayed replica of itself. The temporal spacing of the notches or nulls is inversely proportional to the time delay between the two signals. When an echolocation signal reflects from an irregular surface, the echo will consist of a mixture of reflections from the peaks and reflections from the depressions of the surface. Such an echo would be similar to the echoes that Schmidt (1988) generated electronically and Simmons et al. (1974) recorded from the plates with holes of different depths, as described in Chapter 1. The power spectrum of the FM signal, then, would be represented in the vigor of discharges viewed across each isofrequency contour of the colliculus, and the

94

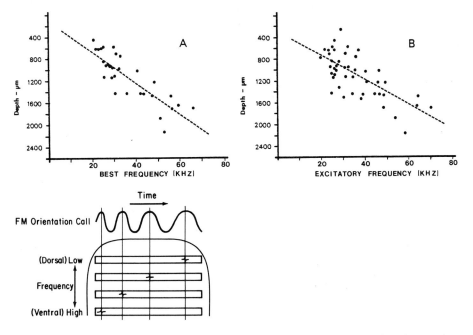

Fig. 4.13. Upper panels Plots of best frequencies versus depth (*A*), and of excitatory frequency versus depth (*B*) in the inferior colliculus of Mexican free-tailed bats. With both measures, there is an orderly arrangement of frequency with depth such that low excitatory and best frequencies are found dorsally, and high excitatory and best frequencies are found ventrally. *Dashed lines* in *A* and *B* are best fits for the data points. (Bodenhamer and Pollak 1981). *Lower panel* Schematic illustration of discharges in time and space within the inferior colliculus, elicited by a downward sweeping FM pulse

accuracy with which each spectral component is represented in its frequency contour is maximized due to the high frequency specificity of the neuronal population for brief FM signals.

4.7.4 Temporal Coding Evoked by Sinusoidally Modulated Signals Related to Target Recognition

As discussed in previous sections, the sharp tuning and overrepresentation of filter units of long CF/FM bats are features conserved in the pathways to the inferior colliculus (Fig. 4.14). As a consequence, the ability to precisely phase-lock to SFM and SAM signals is also evident in collicular filter cells (Schuller 1979a, 1984; Pollak and Schuller 1981; Bodenhamer and Pollak 1983; Reimer 1987) (Fig. 4.15).

Two chief transformations occur between the cochlear nucleus and inferior colliculus with regard to the coding of modulated signals. The first is that phase-locked responses are restricted to lower modulation rates in collicular neurons than in lower auditory nuclei (Vater 1982; Metzner and

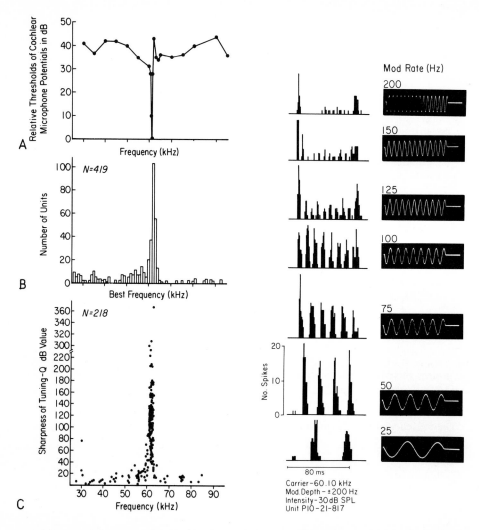

Fig. 4.14 (Figure to the left). A Cochlear microphonic audiogram recorded from one mustache bat showing the sharp sensitive region around 60 kHz. *B* The distribution of best frequencies recorded from single units in the mustache bat's inferior colliculus. There is a pronounced overrepresentation of neurons with best frequencies around 60 kHz, and these frequencies correspond closely to the most sensitive frequency of the cochlear microphonic audiogram. *C* Histogram showing the Q_{10dB} values (tuning sharpness) of single units recorded from the mustache bat's inferior colliculus. The overrepresented 60-kHz neurons have much sharper tuning curves than do units with higher or lower best frequencies. (Pollak et al. 1986)

Fig. 4.15 (Figure to the right). Influence of modulation rates on the phase locked discharges of a filter unit in the mustache bat's inferior colliculus. Duration of SFM signal was 80 ms. The SFM rates are shown to the right of each peri-stimulus time histogram, together with oscillographs of the SFM signals. (Bodenhamer and Pollak 1983)

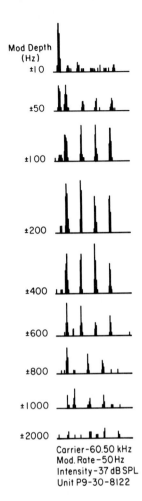

Mod Depth
(Hz)
±10

±50

±100

±200

±400

±600

±800

±1000

±2000

Carrier-60.50 kHz
Mod. Rate-50Hz
Intensity-37 dB SPL
Unit P9-30-8122

Fig. 4.16. Peri-stimulus time historgrams of a filter unit having selectivity, or tuning, for a range of modulation depths. The unit discharged briskly and in tight temporal cadence with the SFM waveform for signals having depths between ± 50 Hz to ± 600 Hz, but discharged most vigorously to depths between ± 100 to ± 400 Hz. (Bodenhamer and Pollak 1983)

Ratdke-Schuller 1987), a difference seen in other mammals as well as in the long CF/FM bats (Møller 1972; Rees and Møller 1983). For example, collicular neurons in both horseshoe and mustache bats phase-lock most readily to SFM signals having modulation rates between about 30–150 Hz (Fig. 4.15) (Schuller 1979 a; Pollak and Schuller 1981; Bodenhamer and Pollak 1983), whereas neurons in the cochlear nucleus commonly lock to modulation rates up to 500 Hz or higher (Vater 1982). The second transformation is that many collicular filter units are selective for the modulation parameters to which they will phase-lock. This property is already present, but is seen less frequently, in lower auditory nuclei (Vater 1982). Many filter units in the inferior colliculus are selective for certain modulation rates, indicative of an insect's wingbeat frequency, or for certain modulation depths, indicative of wingbeat velocity (Pollak and Schuller 1981; Bodenhamer and Pollak 1983). As an illustration, the filter unit in Fig. 4.16 was selective for modulation depth and responded vigorously with phase-locked discharges to

97

SFM signals having modulation depths ranging from ± 50 Hz to ± 400 Hz; it responded to higher or lower modulation depths with a lower discharge rate. Other units responded to non-optimal modulation parameters with lower discharge rates, with a poorer phase-locked precision, or both. Such selectivity for modulation parameters provides the ensemble of filter units with a much richer, and presumably more precise, neural code for an insect's signature (Pollak 1980).

The results presented above were produced with electronically generated SFM and SAM signals, which are only a crude approximation of the more complex signals produced by fluttering targets. It is of some importance, therefore, to show that filter units respond to natural echoes as they do to electronically generated modulated signals. This was elegantly demonstrated by Gerd Schuller (1984). He presented 80-kHz tones to tethered, fluttering insects that were situated at various attitudes with respect to a loudspeaker, and recorded the echoes. He then played these signals to horseshoe bats while recording from filter units in the bat's inferior colliculus. He observed that the neurons responded to the natural echoes in a fashion similar to their responses evoked by SFM and SAM signals. The discharges were in registry with the modulating waveform, but only when the signals had certain modulation parameters.

4.8 The Relevance of Doppler-Shift Compensation

The studies demonstrating the utilization of the CF component for target recognition have also contributed to our understanding of the significance of Doppler-shift compensation for providing the animal with an advantage for hunting insects in acoustically cluttered environments. Doppler compensation is the auditory equivalent of foveation in vision, where moving the eyes keeps images of interest fixated on the fovea (Schuller and Pollak 1979; Pollak et al. 1986). Long CF/FM bats move the frequency of their voices in accord with the Doppler shifts in the echoes they receive. By so doing, the animal ensures itself that the echo from background objects will return at a frequency that will stimulate its "acoustic fovea", the specialized segment of the cochlea which is richly innervated, and thus will excite a large population of filter neurons. When a fluttering insect comes into the bat's acoustic space, the echoes from the insect will have complex acoustic signatures due to the modulations imposed upon the emitted CF component. Insect wingbeats can cause frequency swings of ± 800 Hz, or even larger, around a 60-kHz carrier frequency, and up to ± 2000 Hz around an 80-kHz carrier. Furthermore, bats fly significantly faster than insects, and because the bat compensates only for Doppler shifts imposed by stationary objects in the background (Trappe and Schnitzler 1982), the major portion of compensation for the insect's flight speed has indirectly been achieved. Consequently, it need not compensate for the moment to moment changes of an insect's flight pattern to benefit from its filter neurons. Whether the insect is moving towards or away from the bat, the motion of its wings will produce sufficiently large frequency swings in the echo CF component to evoke phase-locked discharges from at least a portion of the population of filter neurons, and thereby provide the neural substrate for the identification of preferred prey. In the process, it also obtains an advantage for evaluating the location of its target in space, a topic discussed in the following sections.

4.9 Organizational Features Related to Convergence of Inputs at the Inferior Colliculus

A major theme of previous discussions is that each lower brainstem region is tonotopically organized and each has a constellation of defining physiological properties. The projections from each of these regions converge upon comparable isofrequency regions of the inferior colliculus to create a single tonotopic map. In the previous chapter we showed that while there is a substantial convergence of projections on each isofrequency contour, the inputs

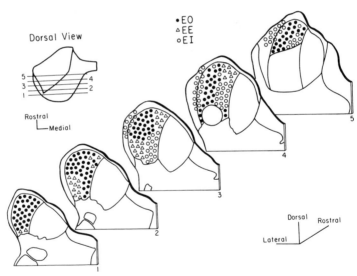

Fig. 4.17. Schematic illustration of the topographic distribution of binaural response classes in the dorsoposterior division of the mustache bat's inferior colliculus. *Filled circles* are monaural cells, *triangles* are E-E neurons and *open circles* are E-I neurons. The region directly below the dorsoposterior division in sections *4* and *5* is the anterolateral division, where frequencies below 60 kHz are represented. The large region medial to the dorsoposterior division in all sections is the medial division, where frequencies higher than the filter frequency are represented. Binaural responses types are shown only for the dorsoposterior division. (Wenstrup et al. 1986)

are not uniformly distributed. The dorsoposterior division of the mustache bat's inferior colliculus, for example, has regional variations in the density of afferents from the various lower nuclei that project to it, so that monaural and binaural regions are topographically arranged within the contour.

As mentioned in the previous chapter, the precise distribution of aural types within the dorsoposterior division has been studied by Wenstrup et al. (1985, 1986). The rationale was that since each lower region is either predominantly monaural or binaural, the degree to which monaural and binaural types are organized within the 60 kHz isofrequency dorsoposterior division could be evaluated physiologically. The three aural types were classified as either monaural (E-O), excitatory-excitatory (E-E) or as excitatory-inhibitory (E-I). By mapping both single and multiunit responses, several zones in the dorsoposterior division could be reliably identified, each having a predominant aural type (Fig. 4.17). Monaural neurons are located along the dorsal and lateral parts of the dorsoposterior division. E-E cells occur in two regions, one ventrolaterally and the other dorsomedially. E-I neurons also have two zones: the main population is in the ventromedial region, and a second population occurs along the very dorsolateral margin of the dorsoposterior division, perhaps extending into the external nucleus of the inferior colliculus.

4.9.1 Inhibitory Thresholds of E-I Neurons Are Topographically Organized Within the Dorsoposterior Division

The population of E-I neurons is of particular interest because they differ in their sensitivities to interaural intensity disparities, and should therefore be important for coding sound location (Wenstrup et al. 1985, 1986; Fuzessery and Pollak 1985; Fuzessery et al. 1985). Supra-threshold sounds delivered to the excitatory (contralateral) ear evoke a discharge rate that is unaffected by low intensity sounds presented simultaneously to the inhibitory (ipsilateral) ear. However, when the ipsilateral intensity reaches a certain level, and thus generates a particular interaural intensity disparity, the discharge rate declines sharply. Then even small increases in ipsilateral intensity will completely inhibit the cell in most cases (Fig. 4.18). Thus each E-I neuron has a steep interaural intensity disparity function and reaches a criterion inhibition at a specific interaural intensity disparity (Wenstrup et al. 1986; 1988a). The criterion is the interaural intensity disparity that produces a 50% reduction in the discharge rate evoked by the excitatory stimulus presented alone.

We shall refer to this interaural intensity disparity as the neuron's inhibitory threshold. An inhibitory threshold has a value of 0 dB if equally intense signals in the two ears elicit the criterion inhibition. An inhibitory threshold is assigned a positive value if the inhibition occurs when the ipsilateral signal is louder than the contralateral signal, and it is assigned a negative value if the intensity at the ipsilateral ear is lower than at the contralateral ear when the discharge rate is reduced by 50%. The inhibitory thresholds of E-I neurons in the dorsoposterior division vary from +30 dB

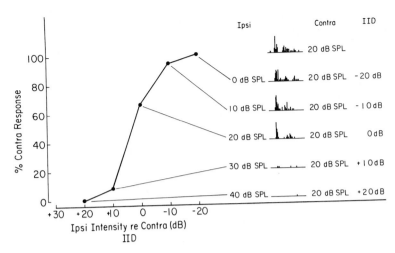

Fig.4.18. Responses of a single E-I neuron as a function of interaural intensity disparity (*IID*). On the *right* are shown the peri-stimulus time histograms of discharges evoked by the contralateral stimulus alone (*top histogram*) and the inhibition produced as the intensity of the ipsilateral (inhibitory) signal was increased. The best frequency of this neuron was 62 kHz, and its threshold was 10 dB SPL. The inhibitory threshold, defined as the IID value at which the discharge rate declined by 50%, was +4 dB (at an ipsilateral intensity 4 db higher than the contralateral intensity)

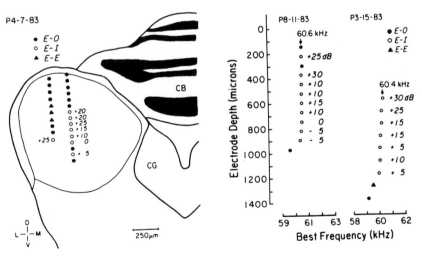

Fig.4.19. Left panel Systematic shift in the inhibitory thresholds of E-I unit clusters shown in transverse section through the mustache bat's inferior colliculus. *Numbers* indicate inhibitory thresholds. All clusters were sharply tuned to frequencies between 63.1–64.0 kHz. *Right panel* Systematic decrease in inhibitory thresholds of E-I neurons with depth in the dorsoposterior division of the inferior colliculus recorded in two dorsoventral penetrations from different bats. (Wenstrup et al. 1985)

to -20 dB, encompassing much of the range of interaural intensity disparities that the bat would experience.

A particularly significant finding is that the inhibitory thresholds are topographically arranged within the ventromedial E-I region of the dorsoposterior division (Fig. 4.19) (Wenstrup et al. 1985, 1986). E-I neurons with high, positive inhibitory thresholds (i.e., neurons requiring a louder ipsilateral stimulus than contralateral stimulus to produce inhibition) are located in the dorsal E-I region. Subsequent E-I responses display a progressive shift to lower inhibitory thresholds. The most ventral E-I neurons have the lowest inhibitory thresholds; they are suppressed by ipsilateral sounds equal to or less intense than the contralateral sounds.

4.9.2 Spatially Selective Properties of 60-kHz E-I Neurons

A neuron's sensitivity to interaural intensity disparities suggests how it will respond to sounds emanating from various spatial locations. To determine more precisely how binaural properties shape a neuron's receptive field, Fuzessery and his colleagues (Fuzessery and Pollak 1984, 1985; Fuzessry et al. 1985; Fuzessery 1986) first determined how interaural intensity disparities of 60-kHz sounds vary around the bat's hemifield. They then evaluated the binaural properties of collicular neurons with loudspeakers inserted into the ear canals, and subsequently determined the spatial properties of the same neurons with free-field stimulation delivered from loudspeakers around the bat's hemifield. With this battery of information, the quantitative aspects of binaural properties could be directly associated with the neuron's spatially selective properties.

The changes in interaural intensity disparity values with azimuthal sound location are shown in Fig. 4.20 (left panel) for 60 kHz in the mustache bat. With regard to azimuth, when the sound emanates from directly in front of the bat, at $0°$ elevation and $0°$ azimuth, equal sound intensities reach both ears, and an interaural intensity disparity of 0 dB is generated. The largest in-

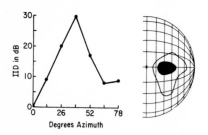

Fig. 4.20. *Left panel* Interaural intensity disparities (*IID*) of 60-kHz sounds for azimuthal locations at 0 degrees elevation. *Right panel* Schematic illustration of IID's of 60 kHz sounds at different azimuths and elevations. Blackened area is region of space where all IID's are 20 dB or greater. The area enclosed by the *solid line* is region of space from which all IID's are 10 dB or greater. (Data from Fuzessery and Pollak 1985)

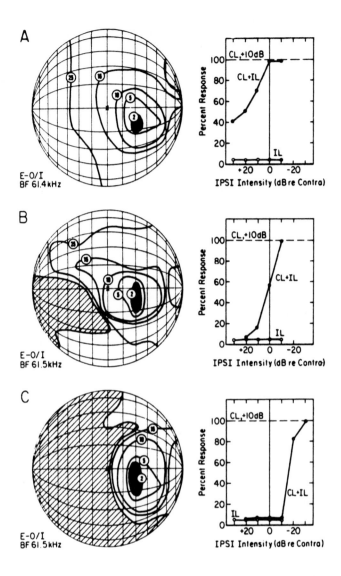

Fig. 4.21 A–C. Spatial selectivity and IID functions for three 60 kHz E-I units in the mustache bat's inferior colliculus. All *lines on globes* connect isothreshold points. Isothreshold contours are drawn in increments of 5 dB. *Blackened area on right* is region of space from which lowest thresholds were elicited. All thresholds in the region differed by at most 2 dB. The uppermost unit (*A*) had a high inhibitory threshold, and could not be completely inhibited regardless of the intensity of the inhibitory sound. The inhibitory threshold of the unit in (*B*) was lower, and the neuron was completely inhibited when the ipsilateral sound was about 10 dB more intense than the contralateral sound. Its spatial selectivity, on the *left,* shows that sounds in portions of the ipsilateral acoustic field, indicated by the *striped area,* were ineffective in firing the unit even with intensities of over 100 dB SPL. Similar arguments apply to the unit in (*C*). The inhibitory threshold of this unit was even lower than for the unit in B. Correspondingly, the region of space from which sounds could not evoke discharges in this unit was also expanded in both the ipsilateral and into the contralateral acoustic field. (Fuzessery and Pollak 1985)

103

teraural intensity disparities originate at about 40° azimuth and 0° elevation, where the sounds are about 30 dB louder in one ear than in the other ear. Within the azimuthal sound field from roughly 40° on either side of the midline, the range of interaural intensity disparities created by the head and ears at 60 kHz is about 60 dB ($+30$ dB to -30 dB). Thus the interaural intensity disparities change on the average by about 0.75 dB/degree. If recordings were made on the left side of the brain and the sound source was located in the left hemifield at 40° azimuth and 0° elevation, the terminology would refer to this as a $+30$ dB interaural intensity disparity. Figure 4.20 (right panel) shows a schematic representation of the interaural intensity disparities generated by 60 kHz sounds in both azimuth and elevation. Notice that an interaural intensity disparity is not uniquely associated with one position in space, but rather a given intensity disparity can be generated by 60 kHz from several spatial locations, a feature that we shall address in a later section.

The binaural properties of three 60 kHz E-I units and their receptive fields are shown in Fig. 4.21. The first noteworthy point is that the spatial position at which each unit had its lowest threshold was the same among all 60 kHz E-I units, at about $-40°$ azimuth and 0° elevation. This location corresponds to the position in space where the largest interaural intensity disparities are generated, i.e., the position in space at which the sound is always most intense in the excitatory ear and least intense in the inhibitory ear. The thresholds increase almost as circular rings away from this area of maximal sensitivity. The highest thresholds are in the ipsilateral sound field, and some units are totally unresponsive to sound emanating from these regions of space.

The second noteworthy point is in that for each 60-kHz E-I unit there is a position along the azimuth that demarcates the region where sounds can evoke discharges from regions where sounds are ineffective in evoking discharges. That azimuthal position of demarcation, and the interaural intensity disparity associated with it, is different for each E-I cell and correlates closely with the neuron's inhibitory threshold. In 60-kHz units having low inhibitory thresholds, those demarcating loci occur along the midline or even in the contralateral sound field, and sounds presented ipsilateral to those loci are incapable of eliciting discharges, even with intensities as high as 110 dB SPL (Fig. 4.21 B and C). Units with higher inhibitory thresholds require a more intense stimulation of the inhibitory ear for complete inhibition, and therefore the demarcating loci of these units are in the ipsilateral sound field. Some units with high inhibitory thresholds (Fig. 4.21 A) could never be completely inhibited with dichotic stimuli. When tested with free-field stimulation these units display high thresholds in the ipsilateral acoustic field.

4.9.3 E-I Neurons Create a Representation of Azimuth

The findings that an E-I neuron's inhibitory threshold determines where along the azimuth the cell becomes unresponsive to sound, coupled with the topographic representation of interaural intensity disparity sensitivities

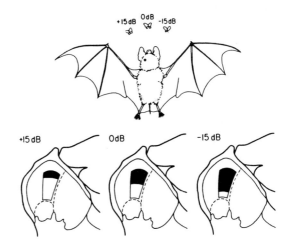

Fig. 4.22. Schematic illustration of the relationship between the value of the interaural intensity disparity produced by a sound source at a given location (moths at *top of figure*) and the pattern of activity in the ventromedial E-I region of the left dorsoposterior division, where IID sensitivities are topographically organized. The activity in this region, indicated by the *blackened area,* spreads ventrally as a sound source moves from the ipsilateral to the contralateral sound field. (Wenstrup et al. 1986)

among E-I cells, have implications for how the *azimuthal* position of a sound is represented in the mustache bat's inferior colliculus. Specifically, these data suggest that the value of an interaural intensity disparity is represented within the dorsoposterior division as a "border" separating a region of discharging cells from a region of inhibited cells (Fig. 4.22) (Wenstrup et al. 1985, 1986, 1988b; Pollak et al. 1986). Consider, for instance, the pattern of activity in the dorsoposterior division on one side generated by a 60-kHz sound that is 15 dB louder in the ipsilateral ear than in the contralateral ear. The interaural intensity disparity in this case is +15 dB. Since neurons with low inhibitory thresholds are situated ventrally, the high relative intensity in the ipsilateral (inhibitory) ear will inhibit all the E-I neurons in ventral portions of the dorsoposterior division. The same sound, however, will not be sufficiently loud to inhibit the E-I neurons in the more dorsal dorsoposterior division, where neurons require a relatively more intense ipsilateral stimulus for inhibition. The topology of inhibitory thresholds and the steep interaural intensity disparity functions of E-I neurons, then, can create a border between excited and inhibited neurons within the ventromedial dorsoposterior division. The locus of the border, in turn, should shift with changing interaural intensity disparity and should therefore shift correspondingly with changing sound location, as shown in Fig. 4.22.

4.9.4 One Group of E-E Neurons Code for Elevation Along the Midline

A population of E-E units has been found that are most sensitive to sounds presented close to or along the vertical midline. However, the elevation at which these cells are most sensitive is determined by the directional properties of the ears for the frequency to which the neuron is tuned (Fig. 4.24, bottom panel) (Fuzessery and Pollak 1985; Fuzessery 1986). Their azimuthal selectivities are shaped by their binaural properties that exhibit either a summation or a facilitation of discharges with binaural stimulation. The contralateral ear is always dominant, having the lowest threshold and evoking the greatest discharge rate. For 60 kHz, the ear is most sensitive, and generates the greatest interaural intensity disparities at $0°$ elevation and about $-40°$ along the azimuth. The position along the azimuth where these cells are maximally sensitive, unlike E-I cells, is not so much a function of ear directionality, but rather is a direct consequence of the interplay between the potency of the excitatory binaural inputs. A sound, for example, presented from $40°$ contralateral will create the greatest intensity at the contralateral ear, due to the directional properties of the ear for 60 kHz. However, as the sound is moved towards the midline, the intensity at the contralateral ear diminishes, but simultaneously, the intensity at the ipsilateral ear increases. Since excitation of the ipsilateral ear facilitates the response of the neuron, the net result is that the response is stronger, and more sensitive, at azimuthal positions closer to, or at the midline than are responses evoked by sounds from the more contralateral positions. In short, many E-E neurons can be thought of as midline units because they are maximally sensitive to positions around $0°$ azimuth. However, the elevation to which they are most sensitive is determined by the directional properties of the ear. Thus 60-kHz E-E units in the mustache bat are always most sensitive at about $0°$ elevation, because the pinna generates the most intense sounds at that elevation (Fig. 4.23).

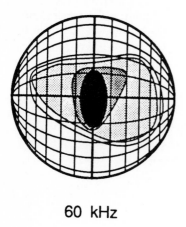

60 kHz

Fig. 4.23. Spatial properties of a 60 kHz E-E neuron that exhibited binaural facilitation. The *blackened* area is the region of space from which the lowest thresholds could be elicited. Isothreshold contours are drawn for threshold increments of 5 dB

106

4.9.5 Spatial Properties of Neurons Tuned to Other Frequencies

The chief difference among neurons tuned to other harmonics of the mustache bat's echolocation calls, at 30 and 90 kHz, is that their spatial properties are expressed in regions of space that differ from 60-kHz neurons, a consequence of the directional properties of the ear for those frequencies (Fig. 4.24 top panel) (Fuzessery and Pollak 1984, 1985; Fuzessery 1986). The binaural processing of neurons tuned to those frequencies, however, are essentially the same as those described for 60-kHz neurons. Moreover, there is even evidence for an orderly representation of interaural intensity disparities in the 90-kHz isofrequency laminae, further supporting the generality of a topology of inhibitory thresholds among E-I cells across isofrequency con-

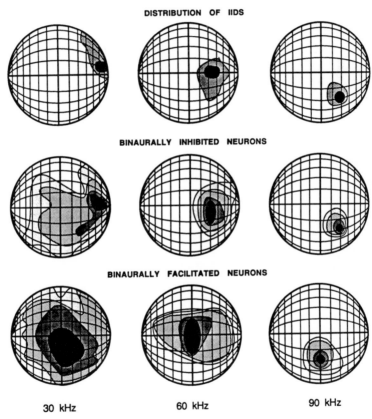

DISTRIBUTION OF IIDS

BINAURALLY INHIBITED NEURONS

BINAURALLY FACILITATED NEURONS

30 kHz   60 kHz   90 kHz

Fig. 4.24. Interaural intensity disparities (IID's) generated by 30, 60, and 90-kHz sounds are shown in *top row*. The *blackened areas in each panel* indicate the spatial locations where IID's are 20 dB or greater, and the *gray areas* indicate those locations where IID's are at least 10 dB. The *panels in the middle row* show the spatial selectivity of three E-I units, one tuned to 30 kHz, one to 60 kHz, and one to 90 kHz. The *blackened areas* indicate the spatial locations where the lowest thresholds were obtained. Isothreshold contours are drawn for threshold increments of 5 dB. The *panels in the bottom row* show the spatial selectivities of three E-E units, tuned to each of the three harmonics. (Fuzessery 1986)

tours (Wenstrup unpublished observations). In short, the sort of binaural processing found in the 60-kHz lamina appears to be representative of binaural processing within other isofrequency contours.

This feature is most readily appreciated by considering the spatial properties of E-I units tuned to 90 kHz, the third harmonic of the mustache bat's orientation calls (Fig. 4.24, middle panel on right). The maximal interaural intensity disparities generated by 90 kHz occur at roughly 40° along the azimuth and $-40°$ in elevation (Fig. 4.24, top panel on right). The 90 kHz E-I neurons, like those tuned to 60 kHz, are most sensitive to sounds presented from the same spatial location at which the maximal interaural intensity disparities are generated. Additonally, the inhibitory thresholds of these neurons determine the azimuthal border defining the region in space from which 90-kHz sounds can evoke discharges from the region where sounds are incapable of evoking discharges (Fuzessery and Pollak 1985; Fuzessery et al. 1985). The population of 90-kHz E-I neurons has a variety of inhibitory thresholds that appear to be topographically arranged within that contour. Therefore a particular interaural intensity disparity will be encoded by a population of 90-kHz E-I cells, having a border separating the inhibited from the excited neurons, in a fashion similar to that shown for 60-kHz lamina. The same argument can be applied to the 30-kHz cells, but in this case the maximal interaural intensity disparity is generated from the very far lateral regions of space (Fig. 4.24, top panel on left).

The spatial behavior of E-E cells tuned to 30 and 90 kHz are likewise similar to those tuned to 60 kHz. The distinction is only in their elevational selectivity, since their azimuthal sensitivities are for sounds around the midline, at 0° azimuth (4.24 bottom panel).

4.9.6 The Representation of Auditory Space in the Mustache Bat's Inferior Colliculus

We can now begin to see how the cues for azimuth and elevation are derived. The directional properties of the ears generate different interaural intensity disparities among frequencies at a particular location. The neural consequence is a specific pattern of activity across each isofrequency contour in the bat's midbrain. Figure 4.25 shows a stylized illustration of the interaural intensity disparities generated by 30, 60, and 90 kHz within the bat's hemifield, and below is shown the loci of borders that would be generated by a biosonar signal containing the three harmonics emanating from different regions of space. Consider first a sound emanating from 40° along the azimuth and 0° elevation. This position creates a maximal interaural intensity disparity at 60 kHz, a lesser interaural intensity disparity at 30 kHz, and an interaural intensity disparity close to 0 dB at 90 kHz. The borders created within each of the isofrequency contours by these interaural intensity disparities are shown in Fig. 4.25 (bottom panel). Next consider the interaural intensity disparities created by the same sound but from a slightly different position in space, at about 45° azimuth and $-20°$ elevation. In this

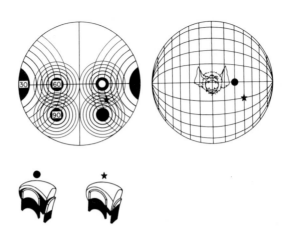

Fig. 4.25. Loci of borders in 30, 60, and 90-kHz E-I regions generated by biosonar signals emanating from two regions of space. The *left panel* is a schematic representation of IID's of 30, 60, and 90 kHz that occur in the mustache bat's acoustic field. The *blackened areas* indicate the regions in space where the maximum IID is generated for each harmonic. The *right panel* depicts the bat's head and spatial positions of two sounds. The *lower panel* shows the borders separating the regions of excited from regions of inhibited neurons in the 30, 60, and 90 kHz isofrequency contours

case there is a decline in the 60 kHz interaural intensity disparity, an increase in the 90 kHz interaural intensity disparity, but the 30 kHz interaural intensity disparity will be the same as it was when the sound emanated from the previous position. The constant interaural intensity disparity at 30 kHz is a crucial point, and it occurs because for a given frequency an interaural intensity disparity is not uniquely associated with one position in space, but rather can be generated from a variety of positions. It is for this reason that the accuracy with which a sound can be localized with one frequency is ambiguous (Blauert 1969/70; Butler 1974; Musicant and Butler 1985; Fuzessery 1986). However, spatial location, in both azimuth and elevation, is rendered unambiguous by the simultaneous comparison of three interaural intensity disparities, because their values in combination are uniquely associated with a spatial location.

This representation becomes ineffective along the vertical midline, at 0° azimuth, where the interaural intensity disparities will be 0 dB at all frequencies and all elevations. The borders among the E-I populations will thus not change with elevation because the interaural intensity disparities remain constant. The population of E-E neurons tuned to different frequencies may be important in this regard (Fuzessery and Pollak 1984, 1985; Fuzessery 1986). E-E units tuned to different frequencies exhibit selectivities for different elevations, and thus as the elevation along the vertical meridian shifts, the strength of responding also shifts for E-E units as a function of their frequency tuning. In combination then, the two major binaural types provide a neural representation of sound located anywhere within the bat's acoustic hemifield.

A striking feature of the above scenario is its similarity, in principle, to the ideas about sound localization proposed previously by Pumphery (1948) and Grinnell and Grinnell (1965). What the recent studies provide are the details of how interaural disparities are encoded and how they are topologically represented in the acoustic midbrain.

As a final point, we raise the possibility that binaural neurons in long CF/FM bats might also have sensitivities to interaural time disparities comparable to those of FM bats. Several features strongly suggest that temporal sensitivity should be a prominent characteristic of the mustache bat's binaural system. The precise temporal synchrony of discharges evoked by modulations in the echo CF component are strikingly similar to phase-locked discharges evoked by low frequency sinusoids in other mammals, and thus are well suited for temporal processing. Also intriguing is the presence of a prominent MSO in the mustache bat, and the dense projection from the MSO to the dorsoposterior division of the inferior colliculus. Since neurons in the MSO of other animals are sensitive to interaural time disparities, it seems reasonable to suppose that the mustache bat's MSO processes binaural signals in a similar manner.

This question has been studied by Wenstrup et al. (1988a), who found that 60 kHz E-I neurons are unaffected by small interaural time disparities. They tested only for sensitivity to time disparities that are generated by the mustache bat's headwidth, around $\pm 50\ \mu s$. However, we point out that binaural neurons sensitive to disparities around $\pm 50\ \mu s$ are rare: Harnischfeger et al., for example, only found 11 such cells in a sample of over 100 neurons, and Pollak (1988) found four cells sensitive to small time disparities out of a sample of over 100 neurons in the inferior colliculus of the Mexican free-tailed bat. The findings of Wenstrup et al. (1988a) do not rule out the possibility that binaural neurons sensitive to very small time variations also occur in long CF-FM bats. However, the weight of evidence suggests binaural E-I cells most likely trade time for intensity in a fashion similar to that discussed previously.

Chapter 5

Processing of Acoustic Information
During Echolocation

5.1 Introduction

In the previous chapter we considered the processing of acoustic information as if the bat were passively listening to sounds reaching its ears. In reality of course echolocation is an active process in which loud pulses are emitted, and the system is then primed or prepared to receive the fainter echoes that return a few milliseconds later. In addition to the type of processing discussed in the previous chapter, there are three chief events that occur with active echolocation. The first concerns mechanisms for reducing the sound pressure of the emitted pulse that actually reaches the inner ear. These mechanisms mainly involve beaming of the emitted pulses and the coordinated action of the middle ear muscles. The second involves central mechanisms whereby the response to an emitted pulse is reduced at higher levels of the auditory system compared with the response evoked when the bat passively listens to the same signal. The third also involves central mechanisms, but in this case the response evoked by a stimulus is qualitatively different during echolocation than it is when the stimulus is presented to an animal that is not echolocating.

5.2 Pulse Beaming and the Action of the Middle Ear Muscles

The directionality of the emitted pulse can affect the sound energy that reaches the ears. Bats emit their orientation sounds either through their mouths, such as the mustache bat and the molossid bats, or beam them through their nostrils, such as the horseshoe bat (e.g., Sales and Pye 1974). Fairly narrow beaming patterns can be achieved through either mechanism (Simmons 1969; Griffin and Hollander 1973; Shimozawa et al. 1974; Schnitzler and Grinnell 1977) so that most of the acoustic energy is directed at targets in front of the bat, with a consequent reduction in the sound level reaching the ear directly. An ancillary point is that stimulation of the cochlea by the emitted signal through bone conduction is also minimized by a loose connection between the cochlea and skeletal elements that are filled with fat deposits or loose connective tissue (Henson 1967).

One of the most potent mechanisms for protecting the cochlea against overstimulation by loud emitted sounds is the action of the middle ear muscles. The ossicular bones in the middle ear are provided with two highly developed muscles: the tensor tympany which attaches to the malleus and the stapedius muscle that is attached to the stapes (Henson 1961, 1967). In 1945 Hartridge proposed that the middle ear muscles of bats contract during pulse emission and maintain the sensitivity of the ear for the perception of echoes. Hartridge's hypothesis was verified in an imaginative experiment by Henson (1965). Henson recorded cochlear microphonic potentials and stapedius-muscle action potentials from Mexican free-tailed bats actively engaged in echolocation. He presented a continuous background tone and monitored the cochlear microphonic potential elicited by the continuous tone. Whenever the bat emitted an orientation call, the amplitude of the cochlear microphonic potential was momentarily reduced due to the contractions of the middle ear muscles. By simultaneously monitoring the emitted pulse with a microphone, he showed that the muscle contraction begins 4–10 ms before the emission of an orientation pulse and reaches a maximum state of contraction just before or at the time of pulse emission. Muscle relaxation begins immediately after maximum tension is achieved, continues over the duration of each pulse and is completed shortly thereafter. Since the stapedius muscle relaxes over the duration of each pulse, the echo energy that reaches the ear when the muscles are relaxed is more efficiently transferred across the middle ear than the preceding pulse energy. In this fashion the muscular contractions prevent overstimulation of the inner ear by attenuating the sound pressure of the emitted pulse by as much as 20 dB. The system is thus in a sensitive state for the perception of faint echoes. This mechanism explains how cochlear microphonic potentials elicited by echoes could be of greater amplitude than those evoked by the more intense outgoing calls (Henson 1965, 1967).

The mechanism outlined above is highly effective for reducing stimulation of the cochlea by signals that are emitted by FM bats. The emitted signals of these bats are so brief that echoes always return to the ears after pulse emission is complete, and consequently these bats never experience pulse-echo overlap. Such a mechanism, however, does not appear well suited for long CF/FM bats, whose emitted pulses are so long that echoes return to the ear while the pulse is still being emitted.

Studies of both mustache and horseshoe bats have shown that their middle ear muscles remain contracted throughout much of the period of pulse emission, and only begin to relax over the duration of the terminal FM component (Henson and Henson 1972; Pollak and Henson 1973; Suga and Jen 1975; Pietsch and Schuller 1987). On the basis of these contraction characteristics, the middle ear muscles of long CF/FM bats should affect the terminal FM component in the same manner as in loud FM bats; that is, the middle ear muscles are contracted and attenuate the emitted FM at the inner ear but are relaxed when the FM component of the echo returns. However, the muscles should also attenuate the heard energy of the emitted CF component and the portion of the echo CF that returns during periods of pulse-echo overlap. Paradoxically, studies in which either cochlear microphonic potentials or evoked potentials from the inferior colliculus were monitored in echolocating bats showed clearly that mustache bats can perceive echoes

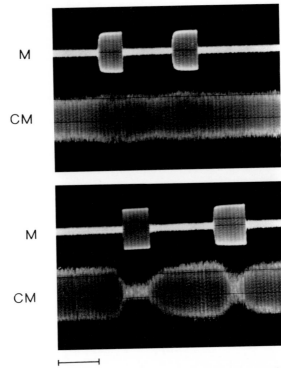

Fig.5.1. The attenuating effect of middle ear muscle contractions on cochlear microphonic (*CM*) potentials during pulse emission in mustache bats. *Upper record* shows the relatively constant amplitude of CM responses to a continuous 61.5 kHz tone at 40 dB above the CM "threshold"; no reduction in amplitude is seen in relation to the emission of the orientation calls. *Lower record* shows the CM response to a 55 kHz tone at 35 dB above CM "threshold"; in this case marked attenuation of the CM potential occurs during pulse emission. The amount of CM attenuation is equivalent to a 30 dB decrease in sound pressure level for the first pulse and about 20 dB for the second pulse. Emitted signals were recorded with a microphone (M). Time scale is 20 ms. (Pollak and Henson 1973)

during periods of pulse-echo overlap when the middle ear muscles are strongly contracted (Henson and Henson 1972; Henson et al. 1982).

Long CF/FM bats overcome the problem of perceiving echoes during periods of pulse-echo overlap because the contracted middle ear muscles act as high pass filters that prevent "low frequencies" from being effectively transmitted through the middle ear yet allow "high frequencies" to be transmitted with little or no attenuation (Pollak and Henson 1973; Pietsch and Schuller 1987). In the mustache bat, for example, frequencies below about 58 kHz are attenuated by as much as 25 dB, while frequencies above 58 kHz are attenuated only slightly or not at all (Fig. 5.1). During most of pulse emission the muscles protect the ear from all components of the 30-kHz fundamental as well as the intense terminal FM component, but they do not appreciably attenuate the 60-kHz CF pulse and echo components. Once the muscles relax at the end of pulse emission, the FM of the echo can be perceived without interference from middle ear muscle activity.

113

Although the high pass characteristics of the middle ear muscles can explain why faint echoes stimulate the inner ear during periods of overlap, these features cannot explain why the loud emitted CF component does not mask the fainter echo that reaches the cochlea while the pulse is still being emitted. There are three features that are seemingly important in this regard. The first is that during flight the frequencies of the pulse and echo CF are substantially different, since these bats compensate for the Doppler shifts in the echoes. Thus there can be a difference in frequency of 1–4 kHz between the emitted and echo CF components. These are substantial frequency differences because the neurons tuned to these frequencies, especially those tuned to the echo frequency, have exceptionally sharp tuning curves. This sharpened frequency selectivity prevents the frequency of the emitted pulse

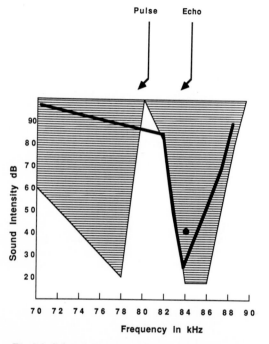

Fig. 5.2. Schematic illustration of an excitatory tuning curve (*solid line*) and the inhibitory surround pattern (*striped area*) in filter units of a horseshoe bat. The tuning curve is determined with tone bursts and shows the range of frequency-intensity combinations capable of eliciting discharges. The inhibitory surround was determined by first setting the tone burst at the best excitatory frequency (84 kHz) at a level about 15 dB above threshold (*black circle*) and then simultaneously presenting a continuous, second tone. The frequency-intensity combinations of the second tone that inhibit the responses to the tone burst define the inhibitory surround. The frequencies of an emitted pulse and Doppler-shifted echo are indicated at the top. Note the pronounced gap in the inhibitory surround region from about 78–82 kHz. Note that if the frequency of the tone burst and continuous tone are 84 kHz, corresponding to the pulse-echo frequencies when echoes are not Doppler-shifted, the unit should be strongly inhibited. However, if the tone burst is 84 kHz and the continuous tone is 4 kHz lower in frequency, which corresponds to the pulse and echo frequencies during Doppler-shift compensation, the continuous tone will evoke little or no inhibition. (After Möller 1978)

114

from exciting the neurons tuned to the frequency of the Doppler shifted echoes (e.g., Suga et al. 1976; Suga and Jen 1977; Möller et al. 1978; Pollak and Bodenhamer 1981).

A second important mechanism is the lateral inhibitory regions of filter neurons that were observed by Möller (1978) in a study of the horseshoe bat's

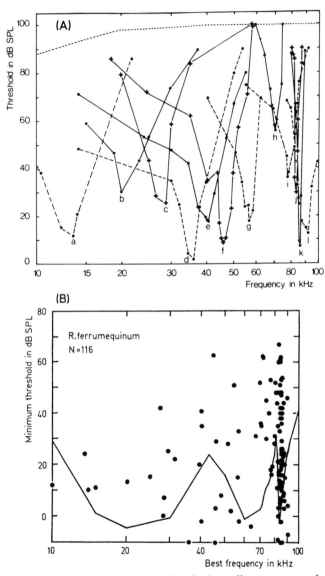

Fig. 5.3. *A* Tuning curves of 12 neurons from the horseshoe bat's auditory nerve and cochlear nucleus. Note the marked threshold increase of units labeled *i* and *h*. *B* Distribution of minimum thresholds as a function of the best frequency from 116 neurons. The *solid line* is the behavioral threshold curve obtained by Long and Schnitzler (1975). (Neural data from Suga et al. 1976)

inferior colliculus. He studied the responses of filter units with two signals presented simultaneously; one signal was a tone burst at the neuron's best excitatory frequency, which elicited discharges when presented by itself, and the second was a continuous tone whose frequency and intensity were varied. Möller found that certain frequencies of the continuous tone inhibited the neuron's response to the tone burst, whereas other frequencies of the continuous tone produced no inhibition, even with high intensities. Thus he observed that collicular filter units had inhibitory surrounds flanking and overlapping with the excitatory tuning curve (Fig. 5.2). The important point is that there is a gap in the flanking inhibitory region just below the frequency at which the bat clamps its echo CF component. The gap extends to frequencies 500–4000 Hz below the neuron's best excitatory frequency. From these data it follows that two signals, an emitted pulse at the resting frequency (the frequency the bat emits when echoes are not Doppler-shifted) and an echo having roughly the same frequency, would create a set of conditions whereby the emitted signal would inhibit the response to the echo. However, the surround inhibitory pattern also suggests that during Doppler compensation, when the frequency of the emitted CF is lowered and falls within the gap of the inhibitory surround, no inhibition would be present. Due to this mechanism, neurons could respond to Doppler-shifted echoes, even during periods of pulse-echo overlap.

The third feature concerns neurons that are tuned to those frequencies of the emitted CF component during Doppler compensation, that are slightly lower than the frequencies of the echo CF. These neurons have substantially higher thresholds than do the neurons sharply tuned to the echo frequency (Neuweiler 1970; Suga et al. 1976) (Fig. 5.3). Behavioral audiograms of horseshoe bats confirm this difference between the threshold for frequencies of the emitted pulse and the threshold for frequencies of the Doppler-shifted echoes (Long and Schnitzler 1975) (Fig. 5.3). Putting these features together, the emitted CF would stimulate a population of neurons having fairly high thresholds, and the echo CF would excite a different neuronal population, whose thresholds are very low and are not inhibited by the frequency of the emitted CF component.

5.3 Central Mechanisms for Attenuating Responses to Emitted Sounds

In addition to pulse beaming and the action of the middle ear muscles, Suga and Schlegel (1972) demonstrated a neural mechanism for suppressing the response to emitted pulses at a higher level of the brainstem auditory pathway. They monitored the evoked potentials from the auditory nerve (N1 response) and lateral lemniscus (LL response) in gray bats (*Myotis grisescens*), a species closely related to little brown bats. Responses from the two auditory areas were evoked by orientation sounds emitted by the bat, and by

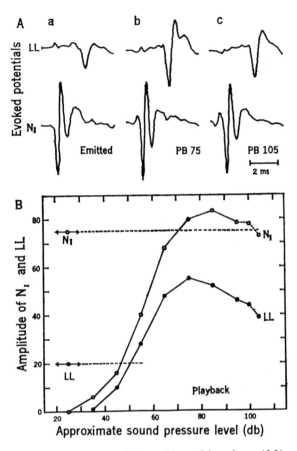

Fig. 5.4. A Summated potentials of the auditory nerve (*N1*) and lateral lemniscus (*LL*). Responses were evoked by sounds emitted by the bat (*a*) and by playback (*PB*) sounds from a tape recorder (*b* and *c*). The amplitudes of the playback sounds were approximately 75 dB SPL for (*b*) and 105 dB SPL for (*c*). The LL response evoked by the bat's emitted call in (*a*) is much smaller than the LL response evoked by the playback calls in (*b*) and (*c*), although the N1 response evoked by the emitted call in (*a*) is nearly the same as the playback N1 in (*b*) and (*c*). *B* The relation between the amplitudes of the playback N1 and LL responses, and the approximate pressure level of the playback sounds. Responses are in arbitrary units. The amplitudes of the N1 and LL evoked by emitted calls are indicated by the *horizontal arrows* and *dashed lines*. The amount of neural attenuation is 20 dB. (Suga and Schlegel 1972)

the same orientation sounds that were played back to the bat through a loud-speaker after being recorded on tape. The salient finding was that the LL response evoked by the bat's emitted call was much smaller than the LL response evoked by the played back signal (Fig. 5.4), even though the N1 responses evoked by the emitted signal and by the played-back signal were of equal amplitude. Thus the responses of lateral lemniscal neurons to the emitted sounds are much smaller than those to sounds that were played back, even when the response of the auditory nerve was the same to both sounds.

117

They also determined the intensity of the played-back sound that evoked an LL response equal to response evoked by the emitted orientation call. In this way they showed that the neural attenuation during pulse emission was equivalent to reducing the signal intensity by as much as 30 dB.

The responses to self-vocalized sounds are attenuated somewhere between the auditory nerve and the lateral lemniscus where the neural attenuating mechanism operates simultaneously with vocalization. The precise origin of the neural attenuation is unknown but most likely occurs somewhere in the superior olivary complex or in the nuclei of the lateral lemniscus. The mechanisms mediating the attenuation of responses to the emitted signals are also unknown but presumably are inhibitory processes of short duration. Suga and Schlegel (1972) suggest that attenuation of the response to the emitted signal, in concert with middle ear muscle activity and other mechanisms mentioned above, protects the auditory system from overstimulation by the emitted pulse and thus allows the auditory system to respond to echoes more effectively than it could have done otherwise.

5.4 Neuronal Response Properties Can Be Modified During Echolocation

A question of considerable importance is whether the coding of acoustic information in actively echolocating bats is the same, or at least similar to that which is evoked in experiments with bats that are listening passively to signals presented to them. Because of the difficulty inherent in recording from single units in behaving bats, only one study, by Gerd Schuller (1979b), has evaluated the response properties of neurons in bats engaged in active echolocation. He recorded the activity of single neurons from the horseshoe bat's inferior colliculus while the bats were emitting echolocation calls and receiving electronically produced echoes played back through a loudspeaker. Some of the neurons that he recorded responded to an emitted pulse and artificial modulated echo in a manner that could be predicted from their response properties to tone bursts and sinusoidally modulated signals presented when the animal was not echolocating. However, many neurons responded to acoustic stimuli entirely differently when the bat was echolocating than when it was listening passively. Of particular interest are the neurons that did not phase-lock to sinusoidally frequency-modulated (SFM) signals, and others that did not respond at all to SFM stimulation in an alert animal listening passively. The same neurons displayed remarkable changes by responding to the modulated signals with phase-locked discharges when the bat was active-ly emitting echolocation calls. Below we illustrate these features with the two neurons in Figs. 5.5 and 5.6.

The unit in Fig. 5.5 responded to emitted calls (VOC in upper left panel) with a tonic firing pattern, but was unresponsive to tone bursts at the same frequency as the emitted CF (AS in upper right panel) and electronically

118

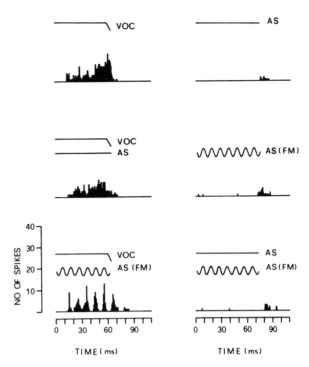

Fig. 5.5. Emission of orientation calls conditions responsiveness of collicular neurons in horseshoe bats. *VOC* indicates discharge pattern of a collicular neuron to vocalized orientation cry alone. *AS* indicates discharges evoked by an artifically generated tone burst at the CF frequency of the echolocation call. (VOC + AS) shows responses to a combination of vocalization and the tone burst. [AS(FM)] indicates responses to a sinusoidally frequency-modulated signal having a center frequency equal to the CF of the emitted call. [VOC + AS(FM)] are responses evoked by a combination of vocalizations and the frequency modulated echo. [AS + (FM)] are responses to combination of tone burst and the frequency modulated echo. See text for further explanation. (Schuller 1979 b)

generated SFM signals [AS(FM) middle right panel]. When the bat emitted orientation calls, and the same SFM signal was simultaneously presented, the unit fired with well defined discharges synchronized to the modulation waveform [VOC + AS(FM) lower left panel]. Control conditions demonstrate that active pulse emission was required to evoke the phase-locked firings. The lower right panel of Fig. 5.5 shows that when a tone burst having the same frequency as the emitted call was presented simultaneously with the SFM signal [AS + AS(FM)], almost no discharges were evoked. Additionally, presenting an 83-kHz tone burst during pulse emission (VOC + AS left middle panel) resulted in a tonic firing pattern, but one in which the discharge pattern was not periodic. These controls show that the necessary conditions to evoke locked firings in this neuron were both active pulse emission and a modulated echo.

A similar phenomenon is shown for another unit in Fig. 5.6. This neuron had a different firing pattern to the CF component of its own emitted call

119

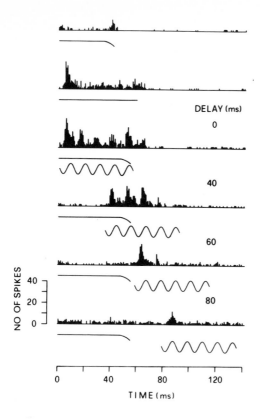

DELAY (ms)

0

40

60

80

NO OF SPIKES

40
20
0

TIME (ms)

Fig. 5.6. Responses of a collicular neuron to emitted orientation calls (*top record*), to tone bursts at the same frequency as the emitted CF component (*second record*), and to combinations of emitted orientation calls and sinusoidal frequency modulated echoes at different time delays from the onset of vocalizations. See text for further explanation. (Schuller 1979 b)

(top panel) than to a tone burst having the same frequency (second panel from top), a characteristic found in about half of the neurons that Schuller sampled. SFM signals presented after pulse emission evoked a predominantly onset response, with no discharges synchronized to the modulation envelope (60 and 80 ms delays in lower panels). The significant feature is that the neuron responded with phase-locked discharges to the SFM signal only during the period of pulse-echo overlap (panels showing 0 and 40 ms delays). In short, the emission of an orientation call opened a gate for this cell, and it was only while the gate was open that the neuron could code for the modulation pattern in the echo CF component with phase locked discharges.

5.5 Conclusions

These studies reveal an intricate coordination of numerous circuits in the nervous system, all acting to optimize the perception of echoes. The beaming of the orientation calls and the action of the middle ear muscles protect the inner ear from overstimulation during emission of the FM component, and circuits activated in synchrony with pulse emission further shield the higher auditory centers from overstimulation by the emitted FM. The nervous system is thereby placed in a highly favorable state to respond to the faint echoes of the FM component that return from nearby objects.

The long CF/FM bats have additional mechanisms for processing the echo CF component. These mechanisms are activated in concert with the emitted calls. The middle ear muscles, in addition to attenuating the FM component, are designed to pass the CF components of both the pulse and echo. Remarkably, the mechanisms at higher levels in long CF/FM bats are apparently designed for two different tasks. Upon pulse emission, the neural circuitry that processes the FM component suppresses the activity of neurons at higher levels of the auditory pathway. However, the circuitry for processing the CF component acts differently, and somehow shapes the discharge pattern in response to its own sounds. Simultaneously, many neurons are primed to code for the modulation pattern of the echo with phase-locked discharges that can be elicited only while the pulse is being emitted. These features could be important for allowing the bat to "focus" on a fluttering target within a limited range, and for discriminating the echoes of its own emitted signals from echoes produced by the signals of other bats. More generally, these studies show that critical neural events occur in a behaving animal and are not present in an alert animal passively listening to sounds.

Chapter 6

Summary and Conclusions

6.1 Specializations of Mustache Bats Reveal Details About the Modular Organization of the Auditory System

Perhaps the major point of the monograph is that the processes seen in the bat's auditory system, whether structural or functional, follow the general mammalian scheme. Our fundamental premise is that even though evolution is a plastic process, it conserves the basic plan, so that all mammalian auditory systems have a common groundplan and employ similar mechanisms. In certain animals, evolution has led to extremely enhanced capacities, such as echolocation in bats. The neural structures associated with this capacity are likewise enhanced. In some cases the enhancement of the neural structures is so pronounced that they yield advantages for experimental analysis.

The central auditory systems of bats are typically mammalian, in terms of both the complement of auditory nuclei and their broad connectional patterns. However, in the details of some parts of the system one finds modifications or elaborations that are clues that these parts have special functional significance. In the mustache bat, for example, the most pronounced modifications are in the cochlea, in the region of the so-called "acoustic fovea". But even here, the only major feature that is truly unique is the resonance of the 60-kHz region of the mustache bat's cochlea. To the best of our knowledge, resonance of this sort has not been previously observed in either intracellular or gross recordings of cochlear microphonic potentials in any mammal.

The "acoustic fovea" represents adaptations that are functionally similar to aspects of the visual and somatosensory systems. In both of these sensory systems the portions of the sensory surface subserving fine discrimination are densely innervated with neurons having small receptive fields so that this part of the sensory surface is greatly magnified in the central nervous system (Hubel and Wiesel 1977; Sur et al. 1980). In long CF/FM bats the cochlear segments representing the acoustic foveae are densely innervated by neurons having exceptionally sharp tuning curves, and this region of the sensory surface is disproportionately represented in all nuclei in the primary auditory pathway.

The analogy with other sensory modalities is even clearer when the integrated actions of both sensory and motor systems are considered. In vision, for example, changing the direction of gaze maintains images of interest fixated on the fovea. Similarly, Doppler compensation ensures that acoustic signals of interest are processed by a region of the sensory surface innervated by a large number of primary fibers having exceptionally narrow tuning curves.

We carry the analogy with the visual and somatosensory systems a step further and point out the modular organization of the mustache bat's auditory system. The concept of a modular organization derives from studies of sensory cortices, in which a module is an elementary unit of neuronal organization that encompasses the total processing from one segment of the sensory surface (e.g., Mountcastle 1978). The most comprehensive example is in the visual system. The studies of Hubel and Wiesel (1977) showed that the apparatus needed to see an area of visual space is contained within a square millimeter of cortex. Thus one can consider an elementary piece of cortex to be a block of tissue that contains neurons with all the possible orientation preferences for one receptive field, and the receptive field is in the same place for each of the two eyes. An adjacent square of cortex would analyze information in the same way for an adjacent part of the visual field, and so on. In parts of the cortex that deal with peripheral field of vision the receptive fields of individual eyes are larger, and thus a small area of cortex deals with a relatively large area of peripheral retina. But the basic organization remains similar to the cortical areas dealing with the fovea, where receptive fields are small. Thus all of the visual cortex is an iterated version of the elementary unit.

In the auditory system, an example of such a modular arrangement is in the inferior colliculus, where each isofrequency contour is an elementary unit. In principle, each collicular contour receives projections from isofrequency regions of the same set of lower nuclei. Moreover, the projections to a particular contour are processed in a way similar to the way that the projections to any other contours are processed. Thus the total processing from a given segment of the cochlear surface receives representation in the corresponding isofrequency contour of the inferior colliculus, and like a cortical module, each contour is basically an iterative version of any other contour.

The most striking module in the mammalian auditory system is the dorsoposterior division of the mustache bat's inferior colliculus, which represents the segment of the cochlear partition that is resonant at 60 kHz. We have stressed earlier that there are features unique to dorsoposterior neurons, such as the sharp tuning curves and sensitivities to periodic frequency modulations. Since these response features are unique, they would seem to contradict the notion of modular identities. However, these properties are not a consequence of central processing but rather are the result of cochlear specializations (Suga et al. 1975, 1976; Suga and Jen 1977; Pollak et al. 1979). The cochlear specializations create the exceptionally sharp tuning curves at 60 kHz. As a consequence of the sharp tuning, stimulation with a modulated 60 kHz signal yields trains of discharges in auditory nerve fibers having temporal cadences that closely follow the temporal features of the modulations in the stimulus. But once the discharges are transmitted to the central auditory pathway, the *processing* of the discharges that emanate from the specialized region of the cochlea appears

124

similar to the processing of discharges that originate in any other portion of the cochlear surface. We say this because connectional patterns, cellular architecture, and synaptic morphology are similar throughout isofrequency regions of each auditory nucleus below the inferior colliculus. In short, there are no anatomical features that distinguish filter neurons from neurons representing other frequencies. The 60-kHz region of each lower nucleus, in turn, projects to the dorsoposterior division (Ross et al. 1988), just as other frequency regions of each lower nucleus project to the other collicular contours. Thus, the general features of the dorsoposterior division are similar to other isofrequency contours in terms of the set of projections it receives (Ross et al. 1988), the cell types present (Zook et al. 1985), and the monaural and binaural response types it contains (Fuzessery and Pollak 1984, 1985; Wenstrup et al. 1986). These common features lead to the conclusion that the dorsoposterior division represents an elementary unit of the inferior colliculus in greatly hypertrophied form.

6.2 Processing of Frequency and Time in the AVCN System

A second theme we have emphasized is the ability of the pathways originating in the AVCN to process spectral and temporal features of acoustic signals. The auditory pathway preserves the frequency to place transformation that originates in the cochlea, where isofrequency regions of lower centers project in a precise fashion to isofrequency regions of higher centers. The extreme version of this partitioning of frequency is in long CF/FM bats, where the sharp tuning curves of filter neurons serve to further demarcate this frequency from other frequency representations. Within any frequency band of the AVCN system, temporal coding is predominant. Temporal coding is expressed most strikingly in two kinds of responses: 1) the phase-locking of filter units to periodic frequency and amplitude-modulated signals and 2) the discharge cadence of constant latency neurons to brief FM signals. Temporal coding is also seen in the ability of binaural neurons to trade time for intensity in bats of the genus Molossus.

The architecture of the AVCN system is well suited for temporal coding. Beginning in the cochlear nucleus, the encoding of frequency modulations in the auditory nerve is preserved by the secure synaptic contacts between auditory nerve fibers and spherical and globular cells of the AVCN. Within the monaural pathways, the temporal information that is preserved in AVCN cells is then transmitted to the nuclei of the lateral lemniscus and inferior colliculus, either directly or with a synaptic interruption in the MNTB, whose axonal collaterals project to the INLL and VNLL. The MNTB, which also is an integral nucleus in the binaural pathway, receives axons from the globular cells of the AVCN. These are some of the largest, and thus fastest, in the mammalian auditory system and synapse upon MNTB neurons with the giant calyces of Held. One of the targets of the monaural pathway, the columnar VNLL, has been discussed in several chapters of this monograph. We have emphasized the columnar alignment of cells in the VNLL and the sequential termination of the endbulbs of Held on each row of cells. These features strongly suggest a mechanism for temporal processing.

In the binaural pathways, the MSO is the nucleus usually associated with processing interaural time disparities, and some of the findings by Harnischfeger et al. (1985) are consistent with this notion. Unfortunately, little additional information exists about the physiological properties of MSO neurons in bats. On the other hand, the elegance of this kind of system is best seen in the avian system. Here, differences in axonal lengths from the two ears generate, within an isofrequency contour, staggered delay lines that converge on MSO neurons. This arrangement is thought to provide the architectural basis for a map of interaural delays (Young and Rubel 1983; Carr 1986; Sullivan and Konishi 1986).

The connections of the lateral superior olive (LSO) might also provide a mechanism for comparing interaural time disparities. The LSO receives ipsilateral input from spherical cells in the AVCN and indirect contralateral input from globular cells in the AVCN via a synapse in the MNTB. We have already emphasized that MNTB is structurally well suited to preserve the timing of discharges. However, as was discussed in Chapter 4, evidence for temporal sensitivity in LSO neurons is controversial. Harnischfeger et al. (1985) found that the ten LSO neurons that they studied were relatively insensitive to interaural time disparities, whereas limited evidence from studies of the cat's LSO indicates that the neurons are temporally sensitive (Caird and Klinke 1983; Yin et al. 1986). The mode of binaural processing by the LSO and elucidating the similarities and differences between it and the MSO are subjects of considerable importance to auditory neuroscience.

6.3 Functional Significance of Monaural and Binaural AVCN Pathways

A third theme of the monograph is that the monaural and binaural pathways that originate in the AVCN are largely distinct; the binaural pathway innervates the superior olives and DNLL, and the monaural pathways innervate the VNLL, INLL or project directly to the inferior colliculus. The MNTB is exceptional in that it is an important contributor to both the binaural and monaural pathways. In addition, we suggest that the functional distinction between the pathways, as seen below the level of the inferior colliculus, relates to the coding of different features of the external world, although the distinction is probably not absolute. Monaural systems are best suited to convey two types of information: 1) the echo spectrum, which the animal interprets for recognition of targets and 2) comparisons of the arrival time of each frequency in the emitted pulse with its counterpart frequency in the echo, information bats use to determine range. Binaural systems, in contrast, are best suited to convey information about sound location.

We suggest the primacy of monaural pathways for coding the echo spectrum because distortions of the coded signal are minimized in these pathways. To illustrate this point, consider how the coding of the echo spectrum may be modified, first in monaural and then in binaural pathways of the AVCN system. Beginning in the AVCN, neurons discharge in a fashion similar if not identical to the discharge pattern of auditory nerve fibers. The first synapses in the monaural AVCN system are either in the MNTB or in the ventral and intermediate nuclei of the lateral lemniscus. The prevalence of calyciform synapses in this circuit suggests that the temporal features of the original code are well preserved within this pathway.

In contrast, binaural pathways distort the coding of the sound spectrum through interactions from the two ears. In binaural systems, the code would be a function not only of the signal spectrum at the tympanum but also of the location of the sound, in that the discharges of some cells would be enhanced and some would be inhibited depending upon the spatial position of the target. This is not to say that such a code could not be interpreted by the bat for characterizing targets. One might devise mechanisms to compensate for or minimize these modifications, but spectral representation is almost certainly more accurately conveyed in monaural systems.

Similar considerations apply to sound localization. Theoretically, monaural pathways could provide information about the general location of a sound – whether it is in the left or right acoustic hemifield and a rough approximation of its elevation – by virtue of the fidelity with which monaural pathways encode the signal spectrum. Monaural localization cues are generated by the pinna. When a broadband sound strikes the pinna, some frequencies are enhanced, some are unaffected and some are attenuated, according to the angular relationship between the position of the pinna and the location of the sound source. Thus, the spectrum of a broad band sound will be modified in a characteristic fashion at the tympanum as a function of its position in space. Fuzessery and Pollak (1984, 1985) have shown that the spatial selectivity of collicular monaural cells, tuned to a particular frequency, closely follows the directionality of the pinna for that frequency. The discharge vigor of monaural neurons viewed across frequency will reflect the signal's spectrum, and thus provides information about sound location. This mechanism, in principle, could account for the ability of humans to monaurally localize broadband sounds (Butler and Flannery 1980; Musicant and Butler 1984).

One potential problem with monaural localization is that the echo spectrum that reaches the bat's ear is affected by numerous variables. Factors that affect the relative energy in the frequencies of the echo include atmospheric attenuation and properties of the insect such as its size and orientation. The spectrum of the echo is then further modified by the directional properties of the pinna as a function of sound location. Thus the bat has no way of knowing to what degree the echo spectrum that it hears is a consequence of location, or is a consequence of these other factors. Under these conditions it is difficult to visualize how the pattern of activity encoded in a monaural pathway could be associated with a unique location in space.

Binaural processing is largely immune from these uncertainties. The binaural system operates on the intensity difference of a particular frequency at the two ears, which is only a function of the sound location. Whether the echo arriving at one ear has more energy at 60 kHz than at 90 kHz is relatively inconsequential. Rather the binaural system evaluates the relative interaural disparity, whether the interaural intensity disparity of 60 kHz is greater or less than the interaural intensity disparity of 90 kHz. Moreover, the synaptic events generated by particular interaural disparities serve to enhance spatial selectivity. The receptive fields of binaural neurons are, therefore, much sharper than those of monaural neurons.

6.4 Convergence of Pathways at the Inferior Colliculus

Another theme is that the parallel pathways in the lower auditory brainstem converge at the inferior colliculus and recombine to form a single tonotopic map. It is at this level that projections from monaural and binaural pathways first converge on limited sectors of isofrequency contours, perhaps even on individual neurons within those contours. The convergence within each frequency contour, however, is not uniform, but rather there is a partial topological parcellation of the various inputs. Thus, a given region of a frequency contour is dominated by inputs from two or three lower nuclei, but that region also receives less dense projections from a limited subset of other nuclei.

One effect of the topographic parcellation of inputs is to separate monaural and binaural neurons within isofrequency contours of the inferior colliculus. This is well illustrated in the dorsoposterior division where there is strict segregation of monaural and binaural neurons. Within the population of binaural cells, the E-E cells are further spatially segregated from the E-I cells.

Although collicular neurons can be classified into the same broad aural categories as they are in lower nuclei (i.e., as monaural, E-E or E-I), they almost surely differ in some important respects from the lower order neurons. The partial convergence of inputs from both monaural and binaural pathways suggests some type of response transformation, whereby the response properties established in lower nuclei are modified to some degree in collicular neurons. We have noted several of these differences. The response to tone bursts changes from mainly tonic discharge patterns in lower nuclei to largely phasic patterns in the inferior colliculus. The response to modulated signals at the colliculus may be an enhanced selectivity for the rate or depth of modulation. But the complexity of the projection circuitry suggests that there are transformations of a more significant nature than those we have so far identified. The population of E-I neurons in the ventromedial 60-kHz

128

region of the mustache bat's inferior colliculus illustrates this point. Our original expectation was that the collicular E-I region receives a strong direct projection from the contralateral LSO, which would account quite simply for the chief binaural properties of the collicular neurons. The connectional pattern, however, turns out to be much more complex than we had suspected. Although there is limited input from the contralateral LSO, the major inputs arise both from DNLLs and from the ipsilateral INLL. If the basic E-I pattern is established in the LSO, why does the circuit pursue such an intricate and complex route before converging on this part of the inferior colliculus? Surely there must be some important transformations between the LSO and collicular E-I region that are presently not understood. The transformations must be complex, involving differences in arrival times of the inputs, the nature of the synapses, whether excitatory or inhibitory, and the relative efficacy of the synaptic events.

Perhaps the key questions in contemporary central auditory physiology are to define more precisely the nature of the transformations that occur between lower nuclei and the inferior colliculus. We need to understand the microcircuitry of the convergence at the inferior colliculus and the neurotransmitters involved. The answers to these questions will provide clues to the broader question of the functional significance of the transformations.

References

Airapetianz E, Konstantinov AI (1974) Echolocation in nature, Part I. JPRS 63328-1, Nat Tech Inf Serv, Springfield

Aitkin LM (1976) Tonotopic organization at higher levels of the auditory pathway. Inter Rev Physiol. Neurophysiology 10:249–279

Allen GM (1939) Bats. Dover, New York

Bateman GC, Vaughan TA (1974) Nightly activities of mormoopid bats. J Mammal 55:45–65

Blauert J (1969/1970) Sound localization in the median plane. Acoustica 22:205–213

Bodenhamer RD, Pollak GD (1981) Time and frequency domain processing in the inferior colliculus of echolocating bats. Hear Res 5:317–335

Bodenhamer RD, Pollak GD (1983) Response characteristics of single units in the inferior colliculus of mustache bats to sinusoidally frequency modulated signals. J Comp Physiol 153:67–79

Bodenhamer RD, Pollak GD, Marsh DS (1979) Coding of fine frequency information by echoranging neurons in the inferior colliculus of the Mexican free-tailed bat. Brain Res 171:530–535

Boudreau JC, Tsuchitani C (1968) Binaural interaction in the cat superior olive S segment. J Neurophysiol 31:445–454

Bourk TR (1976) Electrical response of neural units in the anteroventral cochlear nucleus of the cat. Ph.D. dissertation, Massachusetts Institute of Technology, Cambridge, MA

Brawer JR, Morest DK, Kane EC (1974) The neuronal architecture of the cochlear nucleus of the cat. J Comp Neurol 155:251–300

Bruce CJ, Goldberg ME (1985) Primate frontal eye fields. I. Single neurons discharging before saccades. J Neurophysiol 53:603–635

Bruns V (1976a) Peripheral auditory tuning for fine frequency analysis by the CF-FM bat, *Rhinolophus ferrumequinum*. I. Mechanical specializations of the cochlear. J Comp Physiol 106:77–86

Bruns V (1976b) Peripheral auditory tuning for fine frequency analysis by the CF-FM bat, *Rhinolophus ferrumequinum*. II. Frequency mapping in the cochlea. J Comp Physiol 106:87–97

Bruns V, Goldbach M (1980) Hair cells and tectorial membrane in the cochlea of the greater horseshoe bat. Anat Embryol 161:51–63

Bruns V, Schmieszek E (1980) Cochlear innervation in the greater horseshoe bat: demonstration of an acoustic fovea. Hear Res 3:27–43

Butler RA (1974) Does tonotopy subserve the perceived elevation of a sound? Fed Proc 33:1920–1923

Butler RA, Flannery R (1980) The spatial attributes of stimulus frequency and their role in monaural localization of sound in the horizontal plane. Percept Psychophys 28:449–457

Caird D, Klinke R (1983) Processing of binaural stimuli by cat superior olivary complex neurons. Exp Brain Res 52:385–399

Cajal S, Ramon Y (1911) Histologie du systeme nerveux de l'homme et des vertebrates. Maloine, Paris

Cant N, Morest DK (1979) Organization of the neurons in the anterior division of the anteroventral cochlear nucleus of the cat: Light-microscopic observations. Neuroscience 4:1909–1923

Cant N, Morest DK (1984) The structural basis for stimulus coding in the cochlear nucleus of the cat. In: Berlin CI (ed) Hearing science. College-Hill, San Diego, pp 37–342

Carr CE (1986) Time coding in electric fish and barn owls. In: Wilczynski W, Zakon HH, Pollak GD (eds) Common principles in the neuroethology of acoustic and electric communication. Brain Behav Evol 28:122–134

Casseday JH, Covey E (1987) Central auditory pathways in directional hearing. In: Yost W, Gourevitch G (eds) Directional hearing. Springer, Berlin Heidelberg New York Tokyo, pp 109–145

Casseday JH, Pollak GD (1989) Parallel auditory pathways: I. Structure and connections. In: Nachtigall P (ed) Animal sonar systems. Plenum Press, New York (in press)

Casseday JH, Covey E, Vater M (1988) Connections of the superior olivary complex in the Rufous Horseshoe bat, *Rhinolophus rouxii*. J Comp Neurol 278:313–329

Casseday JH, Covey E, Vater M (1989a) Ascending binaural pathways to the inferior colliculus in the rufous horseshoe bat, *Rhinolophus rouxii*. In: Nachtigall P (ed) Animal sonar systems. Plenum Press, New York

Casseday JH, Kobler JB, Isbey SF, Covey E (1989b) The central acoustic tract in an echolocating bat: an extralemniscal auditory pathway to the thalamus. J Comp Neurol (in press)

Covey E, Casseday JH (1986) Connectional basis for frequency representation in the nuclei of the lateral lemniscus of the bat, *Eptesicus fuscus*. J Neurosci 6:2926–2940

Covey E, Hall WC, Kobler JB (1987) Subcortical connections of the superior colliculus in the mustache bat, *Pteronotus parnellii*. J Comp Neurol 263:179–197

Crowne DP (1983) The frontal eye field and attention. Psychol Bull 93:232–260

Dallos P, Billone MC, Durrant JD, Wang C-Y, Raynor S (1972) Cochlear inner and outer hair cells: Functional differences. Science 177:356–358

David EE Jr, Guttman N, van Bergeijk WA (1959) Binaural interaction of high-frequency complex stimuli. J Acoust Soc Am 31:774–782

Deatherage BH, Hirsh IJ (1959) Auditory localization of clicks. J Acoust Soc Am 31:486–492

Erulkar SD (1972) Comparative aspects of spatial localization of sound. Physiol Rev 52:237–360

Feng AS, Vater M (1985) Functional organization of the cochlear nucleus of rufous horseshoe bats (*Rhinolophus rouxi*): frequencies and internal connections are arranged in slabs. J Comp Neurol 235:529–553

Feng AS, Simmons JA, Kick SA (1978) Echo-detection and target ranging neurons in the auditory system of the bat, *Eptesicus fuscus*. Science 202:645–647

Flannery R, Butler RA (1981) Spectral cues provided by the pinna for monaural localization in the horizontal plane. Percept Psychophys 29:438–444

Friend JH, Suga N, Suthers RA (1966) Neural responses in the inferior colliculus of echolocating bats to artificial orientation sounds and echoes. J Cell Physiol 67:319–332

Frishkopf LS (1964) Excitation and inhibition of primary auditory neurons in the little brown bat. J Acoust Soc Am 36:1016

Fuzessery ZM (1986) Speculations on the role of frequency in sound localization: Brain Behav Evol 28:95–108

Fuzessery ZM, Pollak GD (1984) Neural mechanisms of sound localization in an echolocating bat. Science 225:725–728

Fuzessery ZM, Pollak GD (1985) Determinants of sound location selectivity in bat inferior colliculus: A combined dichotic and free-field stimulation study. J Neurophysiol 54:757–781

Fuzessery ZM, Wenstrup JJ, Pollak GD (1985) A representation of horizontal sound location in the inferior colliculus of the mustache bat (*Pteronotus p. parnellii*). Hear Res 20:85–89

Galambos R (1942a) The avoidance of obstacles by bats: Spallanzani's ideas (1794) and later theories. Isis 34:132–140

Galambos R (1942b) Cochlear potentials elicited from bats by supersonic sounds. J Acoust Soc Am 14:41–49

Galambos R, Griffin DR (1942) Obstacle avoidance by flying bats. J Exp Zool 89:475–490

Goldberg JM, Brown PB (1968) Functional organization of the dog superior olivary complex: an anatomical and electrophysiological study. J Neurophysiol 31:639–656

Goldberg JM, Brown PB (1969) Responses of binaural neurons of dog superior olivary complex to dichotic stimuli: some physiological mechanisms of sound localization. J Neurophysiol 32:613–636

Goldberg, ME, Bushnell MC (1981) Behavioral enhancement of visual responses in monkey cerebral cortex. II Modulation in frontal eye fields specifically related to saccades. J Neurophysiol 46:773–787

Goldman LJ, Henson OW Jr (1977) Prey recognition and selection by the constant frequency bat, *Pteronotus p parnellii*. Behav Ecol Sociobiol 2:411–419

Goldman PS, Nauta WJH (1976) Autoradiographic demonstration of a projection from prefrontal association cortex to the superior colliculus in the rhesus monkey. Brain Res 116:145–149

Griffin DR (1944) Echolocation by blind men and bats. Science 100:589–590

Griffin DR (1958) Listening in the dark. Yale University Press, New Haven

Griffin DR (1967) Discriminative echolocation by bats. In: Busnel R-G (ed) Animal sonar systems. Vol I. Lab Physiol Acoust, Jouy-en-Josas 78, France, p 273

Griffin DR, Galambos R (1941) The sensory basis of obstacle avoidance by flying bats. J Exp Zool 86:481–506

Griffin DR, Hollander P (1973) Directional patterns of bat's orientation sounds. Period Biol 75:3–6

Griffin DR, Dunning DE, Cahlander DA, Webster FA (1962) Correlated orientation sounds and ear movements of horseshoe bats. Nature (Lond) 196:1185–1186

Grinnell AD (1963a) The neurophysiology of audition in bats: Intensity and frequency parameters. J Physiol 167:38–66

Grinnell AD (1963b) The neurophysiology of audition bats: Temporal parameters. J Physiol 167:67–96

Grinnell AD (1963c) The neurophysiology of audition bats: Directional localization and binaural interactions. J Physiol 167:97–113

Grinnell AD (1963d) The neurophysiology of audition bats: Resistance to interference. J Physiol 167:114–127

Grinnell AD (1967) Mechanisms of overcoming interference in echolocating animals. In: Busnel R-G (ed) Animal sonar systems. Vol I. Lab Physiol Acoust, Jouy-en-Josas 78, France, p 451

Grinnell AD (1970) Comparative auditory neurophysiology of neotropical bats employing different echolocation signals. Z Vergl Physiol 68:117–153

Grinnell AD (1973) Neural processing mechanisms in echolocating bats correlated with differences in emitted sounds. J Acoust Soc Am 54:147–156

Grinnell AD, Grinnell VS (1965) Neural correlates of vertical localization by echolocating bats. J Physiol 181:830–851

Grinnell AD, Hagiwara S (1972) Adaptations of the auditory system for echolocation: Studies of New Guinea bats. Z Vergl Physiol 76:41–81

Grinnell AD, Schnitzler H-U (1977) Directional sensitivity of echolocation in the horseshoe bat, *Rhinolophus ferrumequinum*. II. Behavioral directionality of hearing. J Comp Physiol 116:63–76

Guinan JJ, Norris BE, Guinan SS (1972a) Single units in superior olivary complex: I. Responses to sounds and classification based on physiological properties. Int J Neurosci 4:101–120

Guinan JJ, Norris BE, Guinan SS (1972b) Single units in superior olivary complex: II. Locations of unit categories and tonotopic organization. Int J Neurosci 4:147–166

Hafter ER (1984) Spatial hearing and the duplex theory: How viable is the model? In: Edelman GM, Gall WE, Cowan WM (eds) Dynamic aspects of neocortical function. Neurosci Inst Publ, Wiley and Sons, New York

Hall JL (1965) Binaural interaction in the accessory superior-olivary nucleus of the cat. J Acoust Soc Am 37:814–823

Harnischfeger G, Neuweiler G, Schlegel P (1985) Interaural time and intensity coding in the superior olivary complex and inferior colliculus of the echolocating bat, *Molossus ater*. J Neurophysiol 53:89–109

Harrison JM, Feldman ML (1970) Anatomical aspects of the cochlear nucleus and superior olivary complex. In: Neff WD (ed) Contributions to sensory physiology. Vol 4, Academic Press, New York, p 95

Harrison JM, Irving R (1966) Visual and nonvisual auditory systems in mammals. Science 154:738–743

Harrison JM, Warr WB (1962) A study of the cochlear nuclei and the ascending auditory pathways. J Comp Neurol 119:341–380

Hartridge H (1945) Acoustical control in the flight of bats. Nature (Lond) 156:490–494

Henson MM (1978) The basilar membrane of the bat, *Pteronotus parnellii*. Anat Rec 153:143–158

Henson MM, Henson OW Jr, Jenkins DB (1984) The attachment of the spiral ligament to the cochlear wall: anchoring cells and the creation of tension. Hear Res 16:231–242

Henson OW Jr (1961) Some morphological and functional aspects of certain structures of the middle ear in bats and insectivores. Univ Kansas Sci Bull 42:151–255

Henson OW Jr (1965) The activity and function of the middle-ear muscles in echolocating bats. J Physiol 180:871–887

Henson OW Jr (1967) The perception and analysis of biosonar signals by bats. In: Busnel R-G (ed) Animal sonar systems. Vol II. Lab Physiol Acoust, Jouy-en-Josas 78, France, p 949

Henson OW Jr, Henson MM (1972) Middle ear muscle contractions and their relations to pulse and echo-evoked potentials in the bat, *Chilonycteris parnellii*. In: AIBS-NATO Symp Animal Orientation Navigation, Wallops Station, VA, pp 355–363

Henson OW Jr, Henson MM, Kobler JB, Pollak GD (1980) The constant frequency component of the biosonar signals of the bat, *Pteronotus p. parnellii*. In: Busnel R-.G, Fish JF (eds) Animal sonar systems. Plenum Press, New York, p 913

Henson OW Jr, Pollak GD, Kobler, JB, Henson MM, Goldman LJ (1982) Cochlear microphonics elicited by biosonar signals in flying bats, *Pteronotus p. parnellii*. Hear Res 7:127–147

Henson OW Jr, Schuller G, Vater M (1985) A comparative study of the physiological properties of the inner ear in Doppler shift compensating bats (*Rhinolophus rouxi* and *Pteronotus parnellii*). J Comp Physiol 157:587–607

Hubel DH, Wiesel TN (1977) Functional acrrhitecture of macaque monkey visual cortex. Ferrier Lecture, Proc R Soc Lond 198:1–59

Hutson KA, Glendenning KK, Masterton RB (1987) Biochemical basis for the acoustic chiasm? Soc Neurosci Abstr, Vol 13, Part 1, p 548

Illing R-B, Graybiel AM (1985) Convergence of afferents from frontal cortex and substantia nigra onto acetylcholinesterase-rich patches of the cat's superior colliculus. Neuroscience 14:455–482

Jay MF, Sparks DL (1987) Sensorimotor integration in the primate superior colliculus. II. Coordinates of auditory signals. J Neurophysiol 57:35–5

Jeffress LA (1948) A place theory of sound localization. J Comp Physiol Psych 41:35–39

Jones DR, Casseday JH (1979) Projections to laminae in dorsal cochlear nucleus in the tree shrew, *Tupaia glis*. Brain Res 160:131–133

Kiang NY-S, Watanabe T, Thomas EC, Clark LF (1965) Discharge patterns of single fibers in the cat's auditory nerve. Research Monograph No 35, MIT Press, Cambridge, Massachusetts

Kobler JB, Wilson BS, Henson OW Jr, Bishop AL (1985) Echo intensity compensation by echolocating bats. Hear Res 20:99–108

Kobler JB, Isbey SF, Casseday JH (1987) Auditory pathways to the frontal cortex of the mustache bat, *Pteronotus parnellii*. Science 236:824–826

Kössl M, Vater M (1985) The frequency place map of the bat, *Pteronotus parnellii*. J Comp Physiol 157:687–697

Künzle H, Akert K, Wurtz RH (1976) Projection of area 8 (frontal eye field) to superior colliculus in the monkey: An autoradiographic study. Brain Res 117:487–492

134

Leake PM, Zook JM (1985) Demonstration of an acoustic fovea in the mustache bat, *Pteronotus parnellii*. In: Lim DJ (ed) Abstracts of the Eighth Winter Meeting. Res Otolaryngol, p 27

Link A, Marimuthu G, Neuweiler G (1986) Movement as a specific stimulus for prey catching behavior in rhinolophid and hipposiderid bats. J Comp Physiol 159:403–413

Long GR, Schnitzler H-U (1975) Behavioral audiograms from the bat, *Rhinolophus ferrumequinum*. J Comp Physiol 100:211–219

Lorente de Nó R (1981) The primary acoustic nuclei. Raven, New York

Masterton BR, Diamond IT (1973) Hearing: Central neural mechanisms. In: Carterette E, Freedman M (eds) Handbook of perception. Academic Press, New York

Masterton BR, Thompson GC, Bechtold JK, RoBards MJ (1975) Neuroanatomical basis of binaural phase-difference analysis for sound localization: A comparative study. J Comp Physiol Psych 89:379–386

Metzner W, Radtke-Schuller S (1987) The nuclei of the lateral lemniscus in the rufous horseshoe bat, *Rhinolophus rouxi*. J Comp Physiol 160:395–411

Mills AW (1972) Auditory localization. In: Tobias JV (ed) Foundations of modern auditory theory. Voll II. Academic Press, New York, p 303

Möhres FP (1953) Über die Ultraschallorientierung der Hufeisennasen (Chiroptera-Rinolophidae). Z Vergl Physiol 34:547–588

Møller AR (1972) Coding of amplitude and frequency modulated sounds in the cochlear nucleus of the rat. Acta Physiol Scand 86:223–238

Møller AR (1976) Dynamic properties of primary auditory fibers compared with cells in the cochlear nucleus of the rat. Acta Physiol Scand 98:157–167

Möller J (1978) Response characteristics of inferior colliculus neurons of the awake CF-FM bat, *Rhinolophus ferrumequinum:* II. Two tone stimulation. J Comp Physiol 125:227–236

Möller J, Neuweiler G, Zöller H (1978) Response characteristics of inferior colliculus neurons of the awake CF-FM bat, *Rhinolophus ferrumequinum:* I. Single tone stimulation. J Comp Physiol 125:217–226

Montagu G (1809) Cited by Galambos (1942a). Nicholson's J 23:106–116

Morest DK (1968) The growth of synaptic endings in the mammalian brain: A study of the calyces of the trapezoid body. Z Anat Entwicklungs Gesch 127:201–220

Morest DK, Oliver D (1984) The neuronal architecture of the inferior colliculus in the cat: Defining the functional anatomy of the auditory midbrain. J Comp Neurol 222:209–236

Mountcastle V (1978) An organizing principle for cerebral function: the unit module and the distributed system. In: The mindful brain, MIT Press, Cambridge, Massachusetts, p 7

Moushegian G, Jeffress LA (1959) Role of interaural time and intensity differences in the lateralization of low frequency tones. J Acoust Soc Am 31:1441–1445

Moushegian G, Rupert AL, Whitcomb MA (1964) Medial superior olivary unit response patterns to monaural and binaural clicks. J Acoust Soc Am 36:196–202

Moushegian G, Rupert AL, Langford TL (1967) Stimulus coding by medial superior olivary neurons. J Neurophysiol 30:1239–1261

Musicant AD, Butler RA (1984) The psychophysical basis of monaural localization. Hear Res 14:185–190

Nakajima K (1971) The structure of the medial nucleus of the trapezoid body of the bat with special reference to two types of synaptic endings. J Cell Biol 50:121–134

Neuweiler G (1970) Neurophysiologische Untersuchungen zum Echoortungssystem der Großen Hufeisennase, *Rhinolophus ferrumequinum*. J Comp Physiol 67:273–306

Neuweiler G (1983) Echolocation and adaptivity to ecological constraints. In: Huber F, Markl H (eds) Neuroethology and behavioral physiology: roots and growing pains. Springer, Berlin Heidelberg New York Tokyo, p 280

Neuweiler G (1984a) Auditory basis of echolocation in bats. In: Bolis L, Keynes RD, Maddrell SHP (eds) Comparative physiology of sensory systems. Cambridge University Press, Cambridge, p 115

Neuweiler G (1984b) Foraging, echolocation and audition in bats. Naturwissenschaften 71:46–455

Neuweiler G, Möhres FP (1967) The role of spatial memory in orientation. In: Busnel R-G (ed) Animal sonar systems. Vol I. Lab Physiol Acous, Jouy-en-Josas 78, France, p 129

Neuweiler G, Vater M (1977) Response patterns to pure tones of cochlear nucleus neurons in the CF-FM bat, *Rhinolophus ferrumequinum*. J Comp Physiol 115:119–133

Neuweiler G, Bruns V, Schuller G (1980) Ears adapted for the detection of motion, or how echolocating bats have exploited the capacities of the mammalian auditory system. J Acoust Soc Am 68:741–753

Novick A, Vaisnys JR (1964) Echolocation of flying insects by the bat, *Chilonycteris parnellii*. Biol Bull 127:478–488

Oliver DL, Morest DK (1984) The central nucleus of the inferior colliculus in the cat. J Comp Neurol 222:237–264

O'Neill WE, Suga N (1982) Encoding of target range information and its representation in the auditory cortex of the mustached bat. J Neurosci 2:17–24

Osen KK (1969) The intrinsic organization of the cochlear nuclei in the cat. Acta Otolaryngol 67:352–359

Papez JW (1929) Central acoustic tract in cat and man. Anat Rec 42:60

Peff TC, Simmons JA (1971) Horizontal-angle resolution by echolocating bats. J Acoust Soc Am 51:2063–2065

Pfeiffer RR (1966) Anteroventral cochlear nucleus: Waveforms of extracellularly recorded spike potentials. Science 154:667–668

Pfeiffer RR, Kiang NYS (1965) Spike discharge patterns of spontaneous and continuously stimulated activity in the cochlear nucleus of anesthetized cats. Biophys J 5:301–316

Pietsch G, Schuller G (1987) Auditory self-stimulation by vocalization in the CF-FM bat, *Rhinolophus ferrumequinum*. J Comp Physiol 160:635–644

Poljak S (1926) Untersuchungen am Oktavussystem der Säugetiere und an den mit diesem koordinierten motorischen Apparaten des Hirnstammes. J Psychol Neurol 32:170–231

Pollak GD (1980) Organizational and encoding features of single neurons in the inferior colliculus of bats. In: Busnel R-G and Fish JF (eds) Animal sonar systems. Plenum Press, New York, p 549

Pollak GD (1988) Time is traded for intensity in the bat's auditory system. Hear Res 36:107–124

Pollak GD, Bodenhamer RD (1981) Specialized characteristics of single units in the inferior colliculus of mustache bats: Frequency representation, tuning and discharge patterns. J Neurophysiol 46:605–620

Pollak GD, Henson OW Jr (1973) Specialized functional aspects of the middle ear muscles in the bat, *Chilonycteris parnellii*. J Comp Physiol 84:167–174

Pollak GD, Schuller G (1981) Tonotopic organization and encoding features of single units in the inferior colliculus of horseshoe bats: Functional implications for prey identification. J Neurophysiol 45:208–226

Pollak GD, Henson OW Jr, Novick A (1972) Cochlear microphonic audiograms in the pure tone bat, *Chilonycteris parnellii parnellii*. Science 176:66–68

Pollak GD, Marsh DS, Bodenhamer RD, Souther A (1977) Characteristics of phasic on neurons in inferior colliculus of unanesthetized bats with observations relating to mechanisms for echoranging. J Neurophysiol 40:926–942

Pollak GD, Henson OW Jr, Johnson R (1979) Multiple specializations in the peripheral auditory system of the CF-FM bat, *Pteronotus parnellii*. J Comp Physiol 131:255–266

Pollak GD, Wenstrup JJ, Fuzessery ZM (1986) Auditory processing in the mustache bat's inferior colliculus. Trends Neurosci 9:556–561

Pumphery RJ (1947) The sense organs of birds. Ibis 90:171–199

Pye JD, Flinn M, Pye A (1962) Correlated orientation sounds and ear movements of horseshoe bats. Nature (Lond) 196:1186–1188

Ramprashad F, Money KE, Landolt JP, Lauder J (1978) A neuroanatomical study of the cochlea of the little brown bat (*Myotis lucifugus*). J Comp Neurol 178:347–364

136

Rees A, Møller AR (1983) Responses of neurons in the inferior colliculus of the rat to AM and FM tones. Hear Res 10:301–330

Reimer K (1987) Coding of sinusoidally amplitude modulated acoustic stimuli in the inferior colliculus of the rufous horseshoe bat, *Rhinolophus rouxi.* J Comp Physiol

Rhode WS (1985) The use of intracellular techniques in the study of the cochlear nucleus. J Acoust Soc Amer 78:320–327

Rhode WS, Oertel D, Smith PH (1983) Physiological response properties of cells labeled intracellularly with horseradish peroxidase in cat ventral cochlear nucleus. J Comp Neurol 213:448–463

Ross LS, Pollak GD (1989) Differential projections to aural regions in the 60 kHz isofrequency contour of the mustache bat's inferior colliculus. J Neurosci (in press)

Ross LS, Wenstrup JJ, Pollak GD (1986) Differential projections to monaural and binaural regions of one isofrequency contour in the mustache bat's inferior colliculus. Soc Neurosci Abstr 12, Part 2, p 1271

Ross LS, Pollak GD, Zook JM (1988) Origin of ascending projections to an isofrequency region of the mustache bat's inferior colliculus. J Comp Neurol 270:488–505

Rouiller E, Cronin-Schreiber R, Fekete DM, Ryugo DK (1986) The central projections of intracellularly labeled auditory nerve fibers in cats: An analysis of terminal morphology. J Comp Neurol 249:261–278

Sales G, Pye D (1974) Ultrasonic communication by animals. Wiley and Sons, New York

Saint Marie RL, Ostapoff E-M, Morest DK (1988) Cytochemically disparate pathways ascending from the lateral superior olive in the cat: Coding contralateral dominance in the auditory system. In: Lim DJ (ed) Abstracts of the Eleventh Winter Meeting. Assoc Res Otolaryngol, p 164

Schiller PH, True SD, Conway JL (1980) Deficits in eye movements following frontal eyefield and superior collicular ablations. J Neurophysiol 44:1175–1189

Schmidt S (1988) Evidence for a spectral basis of texture perception in bat sonar. Nature (Lond) 331:617–619

Schnitzler H-U (1967) Discrimination of thin wires by flying horseshoe bats (Rhinolophidae). In: Busnel R-G (ed) Animal sonar systems. Vol I. Lab Physiol Acoust, Jouy-en-Josas 78, France, p 69

Schnitzler H-U (1970) Comparison of echolocation behavior in *Rhinolophus ferrumequinum* and *Chilonycteris rubiginosa.* Bijdr Dierkd 40:77–80

Schnitzler H-U, Flieger E (1983) Detection of oscillating target movements by echolocation in the greater horseshoe bat. J Comp Physiol 153:385–391

Schnitzler H-U, Grinnell AD (1977) Directional sensitivity of echolocation in the horseshoe bat, *Rhinolophus ferrumequinum.* I. Directionality of sound emission. J Comp Physiol 116:51–61

Schnitzler H-U, Henson OW Jr (1980) Performance of airborne animal sonar systems I. Microchiroptera. In: Busnel R-G, Fish JF (eds) Animal sonar systems. Plenum Press, New York, p 109

Schnitzler H-U, Ostwald J (1983) Adaptations for the detection of fluttering insects by echolocation in horseshoe bats. In: Ewert J-P, Capranica RR, Ingle DJ (eds) Advances in vertebrate neuroethology. Plenum Press, New York, p 801

Schnitzler H-U, Suga N, Simmons JA (1976) Peripheral auditory tuning for fine frequency analysis by the CF-FM bat, *Rhinolophus ferrumequinum.* J Comp Physiol 106:99–110

Schnitzler H-U, Menne D, Kober R, Heblich K (1983) The acoustical image of fluttering insects in echolocating bats. In: Huber F, Markl H (eds) Neuroethology and behavioral physiology: roots and growing pains. Springer, Berlin Heidelberg New York Tokyo, p 235

Schuller G (1972) Echoortung bei *Rhinolophus ferrumequinum* mit frequenzmodulierten Lauten. J Comp Physiol 139:349–356

Schuller G (1979a) Coding of small sinusoidal frequency and amplitude modulations in the inferior colliculus of the CF-FM bat, *Rhinolophus ferrumequinum.* Exp Brain Res 34:117–132

Schuller G (1979 b) Vocalization influences auditory processing in collicular neurons of the CF-FM bat, *Rhinolophus ferrumequinum*. J Comp Physiol 132:39–46

Schuller G (1984) Natural ultrasonic echoes from wing beating insects are coded by collicular neurons in the CF-FM bat, *Rhinolophus ferrumequinum*. J Comp Physiol 155:121–128

Schuller G, Pollak GD (1979) Disproportionate frequency representation in the inferior colliculus of horseshoe bats: Evidence for an "acoustic fovea". J Comp Physiol 132:47–54

Schuller G, Beuter K, Schnitzler H-U (1974) Response to frequency shifted artificial echoes in the bat, *Rhinolophus ferrumequinum*. J Comp Physiol 89:275–286

Schweizer H (1981) The connections of the inferior colliculus and organization of the brainstem auditory system in the greater horseshoe bat, *Rhinolophus ferrumequinum*. J Comp Neurol 201:25–49

Shimozawa T, Suga N, Hendler P, Schuetze S (1974) Directional sensitivity of echolocation system in bats producing frequency modulated signals. J Exp Biol 60:53–69

Shneiderman A, Henkel C (1987) Banding of lateral superior olivary nucleus afferents in the inferior colliculus: A possible substrate for sensory integration. J Comp Neurol 266:519–534

Simmons JA (1969) Acoustic radiation patterns for the echolocating bats *Chilonycteris rubiginosa* and *Eptesicus fuscus*. J Acoust Soc Am 46:1054–1056

Simmons JA (1971 a) Echolocation in bats: signal processing of echoes for target range. Science 171:925–928

Simmons JA (1971 b) The sonar receiver of the bat. Ann NY Acad Sci 188:161–184

Simmons JA (1973) The resolution of target range by echolocating bats. J Acoust Soc Am 54:157–173

Simmons JA (1974) Response of the Doppler echolocation system in the bat, *Rhinolophus ferrumequinum*. J Acoust Soc Am 56:672–682

Simmons JA (1979) Perception of echo phase in bat sonar. Science 204:1336–1338

Simmons JA, Kick SA (1984) Physiological mechanisms for spatial filtering and image enhancement in the sonar of bats. Ann Rev Physiol 175:599–614

Simmons JA, Lavender WA, Lavender BA, Doroshwo CF, Kiefer SW, Livingston R, Scallet AC, Crowley DE (1974) Target structure and echo spectral discrimination by echolocating bats. Science 186:1130–1132

Simmons JA, Kick SA, Lawrence BD, Hale C, Escudie B (1983) Acuity of horizontal angle discrimination by the echolocating bat, *Eptesicus fuscus*. J Comp Physiol 153:321–330

Sinex DG, Geisler CD (1981) Auditory nerve fiber responses to frequency-modulated tones. Hear Res 4:127–148

Smith PH, Rhode WS (1987) Characterization of HRP-labeled globular bushy cells in the cat anteroventral cochlear nucleus. J Comp Neurol 266:360–375

Stotler WA (1953) An experimental study of the cells and connections of the superior olivary complex of the cat. J Comp Neurol 98:401–:432

Suga N (1964a) Recovery cycles and responses to frequency modulated tone pulses in auditory neurons of echolocating bats. J Physiol 175:50–80

Suga N (1964b) Single unit activity in the cochlear nucleus and inferior colliculus of echolocating bats. J Physiol 172:449–474

Suga N (1967) Discussion of Perception and analysis of biosonar signals by bats. In: Busnel R-G (ed) Animal sonar systems. Vol II. Lab Physiol Acoust, Jouy-en-Josas 78, France, p 1004

Suga N (1970) Echo-ranging neurons in the inferior colliculus of bats. Science 170:449–452

Suga N (1973) Feature extraction in the auditory system of bats. In: Møller AR (ed) Basic mechanisms in hearing. Academic Press, New York, p 675

Suga N (1978) Specialization of the auditory system for reception and processing of species-specific sounds. Fed Proc 37:2342–2354

Suga N (1984) The extent to which biosonar information is represented in the bat auditory cortex. In: Edelman GM, Gall WE, Cowan WM (eds) Dynamic aspects of neocortical function. Wiley and Sons, New York, pp 315–374

Suga N (1989) Parallel-hierarchical processing of biosonar information. In: Nachtigall P (ed) Animal sonar systems. Plenum Press, New York (in press)

Suga N, Jen PH-S (1975) Peripheral control of acoustic signals in the auditory system of echolocating bats. J Exp Biol 69:277–311

Suga N, Jen PH-S (1977) Further studies on the peripheral auditory system of "CF-FM" bats specialized for the fine frequency analysis of Doppler-shifted echoes. J Exp Biol 69:207–232

Suga N, Manabe T (1982) Neural basis of amplitude spectrum representation in auditory cortex of the mustached bat. J Neurophysiol 47:225–255

Suga N, O'Neill WE (1979) Neural axis representing target range in the auditory cortex of the mustache bat. Science 206:351–353

Suga N, O'Neill WE (1980) Auditory processing of echoes: representation of acoustic information from the environment in the bat cerebral cortex. In: Busnel R-G, Fish JF (eds) Animal sonar systems. Plenum Press, New York, p 589–614

Suga N, Schlegel P (1972) Neural attenuation of responses to emitted sounds in echolocating bats. Science 177:82–84

Suga N, Neuweiler G, Möller J (1976) Peripheral auditory tuning for fine frequency analysis by the CF-FM bat, *Rhinolophus ferrumequinum*. IV Properties of peripheral auditory neurons. J Comp Physiol 106:111–125

Suga N, Simmons JA, Jen PH-S (1975) Peripheral specializations for fine frequency analysis of Doppler-shifted echoes in the CF-FM bat, *Pteronotus parnellii*. J Exp Biol 63:161–192

Sullivan WE, Konishi M (1986) A neural map of interaural phase difference in the owl's brainstem. Proc Nat Acad Sci 83:8400–8404

Sun X, Jen PH-S (1987) Pinna position affects the auditory space representation in the inferior colliculus of the FM bat, *Eptesicus fuscus*. Hear Res 27:207–219

Sur M, Merzenich MM, Kass JH (1980) Magnification, receptive field area and "hypercolumn" size in areas 3b and 1 of somatosensory cortex in owl monkeys. J Neurophysiol 44:295–311

Suthers RA, Wenstrup JJ (1987) Behavioral discrimination studies involving prey capture by echolocating bats. In: Fenton MB, Racey P, Rayner JMV (eds) Recent advances in the study of bats. Cambridge University Press, Cambridge, p 122

Trappe M, Schnitzler H-U (1982) Doppler-shift compensation in insect-catching horseshoe bats. Naturwissenschaften 69:193–194

Tsuchitani C (1977) Functional organization of lateral cell groups in the cat superior olivary complex. J Neurophysiol 40:296–318

Tsuchitani C (1982) Discharge patterns of cat lateral superior olivary units to ipsilateral tone-burst stimuli. J Neurophysiol 47:479–500

Tsuchitani C, Boudreau JC (1966) Single unit analysis of cat superior S segment with tonal stimuli. J Neurophysiol 29:684–697

Vater M (1980) Coding of sinusodally frequency-modulated signals by single cochlear nucleus neurons of *Rhinolophus ferrumequinum*. In: Busnel R-G, Fish JF (eds) Animal Sonar System. Plenum Press, New York, p 999

Vater M (1981) Single unit responses to linear frequency-modulations in the inferior colliculus of the greater horseshoe bat, *Rhinolophus ferrumequinum*. J Comp Physiol 141:249–264

Vater M (1982) Single unit responses in the cochlear nucleus of horseshoe bats to sinusoidal frequency and amplitude modulated signals. J Comp Physiol 149:369–388

Vater M (1987) Narrow-band frequency analysis in bats. In: Fenton MB, Racey P, Rayner JMV (eds) Recent advances in the study of bats. Cambridge University Press, Cambridge, p 200

Vater M, Schlegel P (1979) Comparative auditory neurophysiology of the inferior colliculus of two molossid bats, *Molossus ater* and *Molossus molossus*. J Comp Physiol 131:147–160

Vater M, Feng AS, Betz M (1985) An HRP-study of the place-frequency map of the horseshoe bat cochlea: morphological correlates of the sharp tuning to a narrow frequency band. J Comp Physiol 157:671–686

Warr WB (1982) Parallel ascending pathways from the cochlear nucleus: Neuroanatomical evidence of functional specializations. In: Neff WD (ed) Contributions to sensory physiology. Vol 7, Academic Press, New York, p 1–38

Webster FA, Durlach NI (1963) Echolocation systems of the bat. MIT Lincoln Lab Rep No 41-G-3, Lexington

Webster FA, Griffin DR (1962) The role of the flight membranes in insect capture in bats. Anim Behav 10:332–340

Wenstrup JJ, Ross LS, Pollak GD (1985) A functional organization of binaural responses in the inferior colliculus. Hear Res 17:191–195

Wenstrup JJ, Fuzessery ZM, Pollak GD (1986) Binaural response organization within a frequency-band representation of the inferior colliculus: Implications for sound localization. J Neurosci 6:692–973

Wenstrup JJ, Fuzessery ZM, Pollak GD (1988a) Binaural neurons in the mustache bat's inferior colliculus: Responses of 60 kHz EI units to dichotic sound stimulation. J Neurophysiol 60:1369–1383

Wenstrup JJ, Fuzessery ZM, Pollak GD (1988b) Binaural neurons in the mustache bat's inferior colliculus: Determinants of spatial responses among 60 kHz EI units. J Neurophysiol 60:1384–1404

Whitworth RH, Jeffress LA (1961) Time vs intensity in the localization of tones. J Acoust Soc Am 33:925–929

Wilson JP, Bruns V (1983) Basilar membrane tuning properties in the specialized cochlea of the CF-FM bat, *Rhinolophus ferrumequinum*. Hear Res 9:15–35

Yin TCT, Hirsch JA, Chan JCK (1985) Responses in the cat's superior colliculus to acoustic stimuli; II. A model of interaural intensity sensitivity. J Neurophysiol 53:746–758

Yin TCT, Chan JCK, Carney LH (1986) Neural mechanisms of interaural time sensitivity in the cat's auditory brainstem. In: IUPS Satellite Symp Hear Univ California, San Francisco, CA, p 61

Young SR, Rubel EW (1983) Frequency-specific projections of individual neurons in chick brainstem auditory nuclei. J Neurosci 3:1373–1378

Zook JM, Casseday JH (1980) Identification of auditory centers in lower brainstem of two species of echolocating bats: Evidence from injection of horseradish peroxidase into the inferior colliculus. In: Wilson DE, Gardner AL (eds) Proc 5th Int Bat Res Conf, Texas Tech Press, Lubbock, pp 51–60

Zook JM, Casseday JH (1982a) Cytoarchitecture of auditory system in lower brainstem of the mustache bat, *Pteronotus parnellii*. J Comp Neurol 207:1–13

Zook JM, Casseday JH (1982b) Origin ascending projection to inferior colliculus in the mustache bat, *Pteronotus parnellii*. J Comp Neurol 207:14–28

Zook JM, Casseday JH (1985) Projections from the chochlear nuclei in the mustache bat, *Pteronotus parnellii*. J Comp Neurol 237:307–324

Zook JM, Casseday JH (1987) Convergence of ascending pathways at the inferior colliculus of the mustache bat, *Pteronotus parnellii*. J Comp Neurol 261:347–361

Zook JM, DiCaprio RA (1988a) Projections of the medial nucleus of the trapezoid body to the lateral superior olive: An in vitro morphological study. In: Lim DJ (ed) Abstracts of the Eleventh Winter Meeting. Assoc Res Otolaryngol, p 165

Zook JM, DiCaprio RA (1988b) Intracellular labeling of afferents to the lateral superior olive in the bat *Eptesicus fuscus*. Hear Res 34:141–148

Zook JM, Leake PA (1988) Correlation of cochlear morphology specializations with frequency representation in the cochlear nucleus and superior olive of the mustache bat, *Pteronotus parnellii*. J Comp Neurol (in press)

Zook JM, Winer JA, Pollak GD, Bodenhamer RD (1985) Topology of the central nucleus of the mustache bat's inferior colliculus: Correlation of single unit properties and neuronal architecture. J Comp Neurol 231:530–546

Subject Index

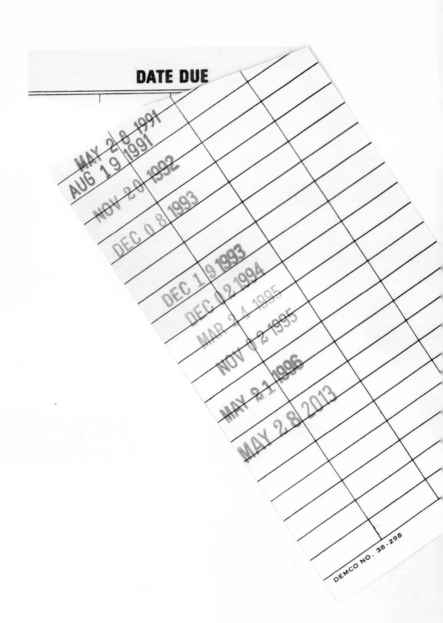

DATE DUE

MAY 2 8 1991
AUG 1 9 1991
NOV 2 0 1992
DEC 0 8 1993
DEC 1 9 1993
DEC 0 2 1994
MAR 2 4 1995
NOV 0 2 1995
MAY 2 1 1996
MAY 2 8 2013